BIG CATS IN BRITAIN
YEARBOOK
2008

Edited by Corinna and Jonathan Downes
Cover and internal design by OranjKat for CFZ Communications
Using Microsoft Word 2000, Microsoft , Publisher 2000, Adobe Photoshop CS.

First published in Great Britain by CFZ Press

CFZ Press
Myrtle Cottage
Woolfardisworthy
Bideford
North Devon
EX39 5QR

© CFZ MMVIII

All rights reserved. Without limiting the rights under copyright reserved above, no part of this publication may be reproduced, stored in or introduced into a retrieval system, or transmitted, in any form of by any means (electronic, mechanical, photocopying, recording or otherwise), without the prior written permission of both the copyright owners and the publishers of this book.

ISBN: 978-1-905723-23-2

Contents

- 05 Forward by Merrily Harpur.
- 07 Big Cats in Britain Conference by Christine Hall.
- 13 Mixing Art and the Curriculum by Brian Percival.
- 19 Every Village Should Have One by John Beart.
- 23 "If There Were Big Cats, I'd See Them!" by Rick Minter.
- 31 Territory or Range by Christopher Johnston.
- 35 Big Cat Killed on Bypass by Frank Turnbridge.
- 41 Bringing Big Cats into the Work Place by Chris Hall.
- 45 Big Cat Diary - England.
- 181 Big Cat Diary - Scotland.
- 219 Big Cat Diary - Wales.
- 226 Big Cat Diary - N Ireland.
- 229 Big Cat Diary - Ireland.
- 231 Big Cat Diary - USA.
- 239 Big Cats in Britain by Mark Fraser.
- 246 Credits.
- 247 Websites.

•

A special thank you to the following: (in order of appearance in the yearbook) Merrily Harpur, Christine Hall, Brian Percival, Rick Minter, John Beart, Jan Williams, Chris Hall, Rachel Lacey, Chris Moiser, Mark Williams, Alan White, Alan Fleming, Christopher Johnston, Anthony Bevan, Darren Naish, Paul Westwood, Frank Turnbridge, Nicole Webb, Cheryl Hudson, Neil Hughes, Colin Mayes, Elizabeth Andrews, Nigel Spencer, Rob Cave, Donna Brown, David McQuirk, Ian Bond, Martin Rainer, Marcus Mathews, Terry Dye, Chris Mullins, Geoff Featherstone , Katy Jordan, Graham Levy, Di Francis, Sandy Smith, Shaun Stevens, Mark Fraser, Mark Maylin, Mark Martin, Charlie McGuinness, Alex Mistretta, John Finch, Hannah Fraser and Loren Coleman.

Foreword

by Merrily Harpur

This is the third *Big Cats in Britain* Yearbook to be inspired and compiled by the group's indefatigable leader Mark Fraser, and once again it details an extraordinary number of sightings of big cats throughout Britain.

Every county of England had its mysterious felines in 2007, with the exception of London - though there were plenty of sightings in the home counties. Yorkshire produced the highest number of reports - 64 - almost double those of the runner-up, Devon, with 38. Somerset had 37 and Leicestershire almost as many with 36, while Gloucestershire and Wiltshire had 31 reported sightings apiece. In Scotland Fife and Aberdeenshire topped the league with 36 sightings between them, though Wales had a quiet year with only six and four reports coming from Glamorgan and Denbighshire respectively. Co. Monaghan in Northern Ireland again had a spate of sightings - six reported, and there were three from northern counties of the Republic of Ireland, Donegal and Sligo.

In short there were 675 apparent sightings logged in 2007, but behind this statistic are 675 witnesses, each - in their own words - 'shocked' or 'stunned', amazed, interested or curious enough to tell someone of their experience. Behind them again are hundreds more witnesses who have not mentioned their glimpse of a big cat to anyone, whether out of fear of being disbelieved or simply because they do not know who to tell. Mark Fraser's national research network, *Big Cats*

BIG CAT YEAR BOOK 2008

In Britain or BCIB has been providing that service for many years, each sighting being acknowledged, logged, published on the members' area of the website, and if necessary followed up by a local member.

BCIB members are a diverse bunch, firmly bonded by their fascination with the mystery of Britain's many panther, puma or lynx-like big cats, and their determination to find out more about their nature and provenance. Some have added their thoughts on and theories about these knotty questions to this volume, along with other writers who reflect on their own experience of seeing a big cat or describe their local reports.

In short this is the most indispensable guide to latest sightings and developments in research into the most strange and secret denizens of our landscape. I defy anyone to read it and not find - the next time they go for a walk - their view of the countryside, the home we share with these elusive creatures, subtly changed.

Merrily Harpur,
January 2008.

Big Cats in Britain Conference 2007

by Christine Hall

The conference, which took place in March 2007, was an event of importance. It continued the forward momentum of the 2006 conference held in Marston Trussell, which was organised by Merrily Harpur. This marked the start of the recognition of the big cat issue as an important subject to be taken seriously by the public and the media. The 2007 conference came into being from the vision of Mark Fraser who, for many years, has been a committed and passionate researcher who has always been willing to share his knowledge and experience with others. He is particularly interested in those members of the public who have a sighting feeling able to come to an organisation were they will be taken seriously. Hence the conference and the BCIB.

In this article, I would like to look at the value of the conference rather than its actual content, as others (see Dr. Darren Naish available on the BCIB website) have written about this. The conference was important because it brought together some of the foremost researchers, and thinkers, in the world of big cat debate. These researchers brought gravitas to an issue not previously taken seriously by the public and media. However, times are changing, and this is thanks to the hard work and sharing of information that occurs between these particular researchers who have been brought together through the efforts of Mark and Hannah Fraser who organised the conference. As the founder of the BCIB, it is due, in no small part, to Mark's willingness to share his experience and information, which has attracted other like-minded people to the BCIB. As Nigel Brierly (BCIB Hon. Life President) said in his forward to the 2006 year book: *"Mark formed the Big Cats in Britain research group just at the time it was needed"*. I, for one, concur with that as I know many others do.

Left to right, back row: *Jonathan Downes, Alan White, Colin Mayes, Dave Mitchel, Darren Naish, Shaun Stevens, Rick Minter, Terry Dye.* **Centre:** *Mark North & Corinna Downes* **Bottom:** *Marcus Mathews, Christopher Johnston, Mark Martin, Chris Moiser, Charlie & Daargh McGuinness.* **Front:** *Mark Fraser* **Inset:** *Di Francis and the Revack Cat.*

Among the researchers present was Di Francis, who is widely known for being one of the first people to explore the big cat issue and to carry out studies. She has, over the years, influenced many other researchers with her knowledge, enthusiasm and her books, as well as her ability to bring structure and passion to the subject. We were very honoured to have her as a speaker, and she certainly remained focused on the subject of British Big Cats. We were very pleased to welcome the other researchers and contributors, such as Dr. Darren Naish, Jon McGowan, Terry Dye, Jon Downes, Marcus Matthews, Chris Hall, Alan White, David Mitchell, Shaun Stevens, and Rick Minter, who was a great MC for the weekend, Chris Moiser, Charlie McGuiness, Chris Johnson, Mark Martin, and Mark North to mention just a few! All these people, and the rest who attended, made it a great and valuable conference.

I believe one of the main values, which this conference highlighted, was in bringing these people together in order to widen debate and present new ideas. People commented that it was good to actually meet people they had only communicated with on the website, or telephone. The level of cohesiveness with which everyone was willing to share, and discuss their ideas and research, was good to see. It is only when groups like this share effort that things move forward.

Traditionally media interest tends to accentuate the 'dramatic', as that is often what catches the eye of the reader. However, over the past decade or so this has begun to change with a more serious attempt to present a clearer picture to the reader. This is highlighted in the way the Ireland tour was filmed, and also the 'Heart of the Country' series and 'Animal 24/7', and others. Further value lay in the presence of the various media, which Mark had organised for this occasion. Endomol in particular, who are making a documentary, were present for the

Mark Fraser and the Revack Cat skin, on loan from Edinburgh Natural History Museum courtesy of Dr Andrew Kitchener

whole weekend. The interviewer seemed to have a serious and interested approach to the issue. There had been a lot of local newspaper interest in the form of articles and interviews. Although this did not generate the numbers of general public attending, it did, however, put the BCIB out there in the public domain to let those who have the sightings know there is an organisation for them. Also from the publicity, we gained a local representative.

The key gains, which I gathered from my talks with people at the conference and after, were that it was good to have that level of communication, for example, when Di Francis talked about her theory of the indigenous cat. There were those who agreed, but equally there were others who were not convinced, and would like to see a lot more evidence. Therefore, a healthy balanced debate took place. This level of communication is what helps a group to grow and develop, and come up with answers eventually, or if not conclusive answers, at least to set a template for the future.

The latest data shared by the key researchers was very much appreciated by people, as was the openness with which speakers answered the audience's questions. Some people have mentioned that this research can be quite isolating so to be able to get together in this way re-enforced being part of an organisation, which is so valuable to many people because it aids motivation and enthusiasm.

BCIB broke the mould when it came to the sharing of knowledge and experience rather than keeping it encapsulated, which does not create progress. This conference really highlighted this.

A further valuable aspect of the conference was the recognition of the hard work and effort, which the researchers undertake, and this recognition and appreciation was shown by a presentation to Shaun Stevens. This will be carried forward to the 2008 conference with another award to a worthy recipient.

The conference also highlighted the value of having members with a scientific background. We live in a society, which is sceptical of any study, which does not come from an 'expert'. We saw how those with scientific backgrounds bring a certain rigour and high level of expectation to their subject, how they demand and search for good evidence. This flags up the value of a group with researchers from different backgrounds, and heightens the quality of the research, debate and evidence presented.

On a lighter note it was not all serious talk and debate - there was lots of fun too. A quiz was held, and various other group bonding exercises. (The bar!) In fact, there was quite a buzz to the whole weekend, with interviews and discussion groups taking place.

In conclusion, the conference was very successful in that it met many of its aims, one of which was to bring key researchers together for debate, and sharing of views and experience. It brought gravitas to a subject often perceived as fringe, although it was clearly demonstrated by the speakers that the issue of big cats in our countryside is a very serious issue with many ramifications.

Mixing Art and the Curriculum
by Brian Percival

Twenty-five years ago, I had to do a sponsored walk when I was at school. The ten-mile walk was along a disused railway track in Wolverhampton. *"Great"* I was thinking, *"how can I get out of this?"* Then something happened to change that. The local press reported that somebody had seen a big cat on the loose near Wolverhampton. Guess where! Wow – this walk might be a bit more interesting.

As it happens, I never saw a big cat, and I remember being worn out (from looking over my shoulder?) but I am still mildly obsessed with them! To me, the prospect of UK resident big cats (however they got here, or whatever they are) prowling our lanes and fields gives the countryside an edge. In a countryside increasingly managed and encroached upon by man, it seems to me that we need that element of mystery.

I am a professional artist based in the North West. I also work at the University of Salford. I have worked on a number of art projects, which have had a Fortean flavour to them. These have included 'The Chapel Street Ghost Walk' in Salford, experiment in EVP at Salford University, and experiments with ESP at Newton Street Gallery in Manchester (with occasional collaborator Lee Patterson). I have also done other small site-specific experiments based around UK landscape and wildlife.

In late 2005, I was approached by Bolton Museum and Art Gallery to come up with some ideas for an educational art project called 'Connecting with Collections'. The project links an artist with a school, and part of the Bolton Museum collection. I, and the project manager Denise Bowler, would then devise an art project for one class of Year 5s (9 to 10 year olds) which would run over a 12-week perio, the idea being that myself and the children would use something from the collection as a basis for artwork in school.

The reason I was asked to do this was because previous Bolton Museum projects had tended to be a little bit traditional. There is no problem with this – they simply wanted a change. My work is what is termed process and performance based. I treat every step of the project as an artwork in itself. Each child has a role within the group, and each week the whole class would be doing an activity – leading to an exhibition at the museum.

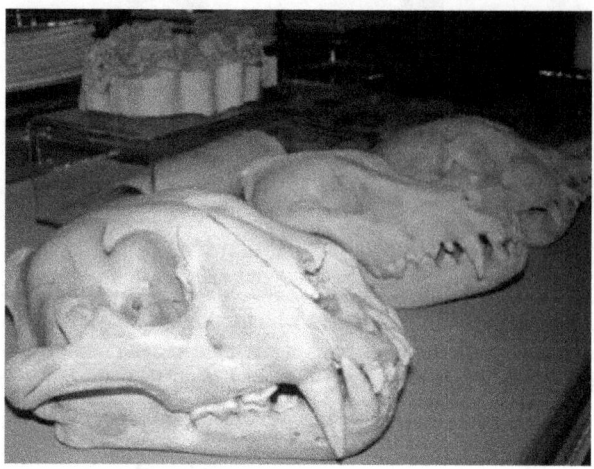

The first thing I needed to do was have a look at the museum collections and archives. My first port of call was the Natural History Department, and curator Pat Francis. *"Have you got anything unusual?"* I asked. *"Well we used to have a two headed pig but we swapped it with another museum for a mermaid."* This was a good start (as was the fake meteor) but I was not sure I could convince a school to work with the theme!

As it happens, Pat showed me a cast of a paw print and a newspaper article from a big cat sighting in 1996 at Belmont, on the outskirts of Bolton. This was actually on display at the museum for a time, but is now in the archives. According to the witnesses, the cat was black and larger than a domestic. Police were called, a cast taken, and the incident logged. This was to be the starting point of the art project.

The school we were working with was St Stephens and All Martyrs Primary, which is very close to Bolton town centre. There is a surprising amount of countryside close to Bolton - the town is situated on the edge of the Pennines. Merlin, peregrine and buzzards are plentiful, and roe deer can even be seen in the town centre at night. With the school bordering the River Tonge, and large amounts of woodland, the project seemed to fit.

Over the weeks, the class gathered information on local wildlife, habitat, pet ownership and the sighting in Belmont, as well as other NW sightings. We contacted Danny Bamping (Bolton born and bred) who gave us some details of other Bolton big cat sightings. The children used maps and internet research, as well as securing the loan of taxidermy specimens and skulls from the Museum, in order to set up the class as an 'incident room'. We visited the woods, finding both deer and mink tracks, and a hibernating toad! We also collected objects as well as filming, drawing, mapping and photographing the area. Each child had a specific role within a group, and the groups then informed the project as a whole. Though we found no evidence of big cats, the purpose of the project was to engage all children in art work by not taking a traditional painting and drawing approach. By using research, digital AV and collecting as the main tools, it meant all of the class had a role, whatever their perceived artistic ability. Of course, having big cats potentially roaming the countryside is exciting! In addition, children learn more when a subject excites them.

We interviewed school staff, one of whom had seen a big cat around Bolton in the late 90s. Other teachers were sceptical – in fact typical of society in general. It seems that unless you have actually had some kind of encounter, then you do not believe it.

As luck would have it (and is it me or are there many of these coincidences around Fortean subjects?) our project fell right in the middle of a series of Bolton big cat incidents - some within the area we were working. There was a sighting from Julie Foster, a district nurse who was riding a horse in Darcy Lever and spotted a large all black cat (see 2007 yearbook). This led to several articles and letters in the *Bolton Evening News*. I was interviewed by *Channel M News* to give my opinion as to whether a big cat could survive in the countryside, and to show the Belmont cast. For those of you researching, I must stress that this was not set up and was independent of our art project.

The final exhibition of work, sited at Bolton Museum in summer 2006, consisted of a video piece, as well as a large display cabinet. The cabinet featured research, natural history specimens, equipment used, small drawings and maps – in fact pretty much everything we used, recorded and found out over the duration of the project.

The same class, and Bolton Museum, followed this work with a project late in 2006 that documented the local woodland flora and fauna. Called 'In the footsteps of Darwin' the project won the prestigious Rolls Royce Prize for Science in schools.

The second project I was involved in was in January 2007. I was one of two artists selected to deliver a project entitled 'Alchemy: Enquire linking Manchester Museum and the teaching of citizenship within local schools' (St Ambrose Barlow Secondary Salford in my case).

I concentrated on the theme of native, and non-native, flora and fauna. Manchester Museum has a display based on this, and there were many themes within this subject that could be taught in citizenship lessons (though I had to do some research on exactly WHAT citizenship in schools was!).

The work we did was similar to that undertaken in Bolton, but with Year 8 pupils (13 – 14 year olds). The project was very much a collaboration between myself, and class teacher Dawn Hayes (who informed me of a rogue population of chinchillas in Salford!) We had more of an emphasis on the local Salford / Swinton area. This meant a visit to Clifton Country Park, on the banks of the River Irwell, where pupils found out that a dead python had been discovered a few weeks previously after a heavy frost (pets are still being let loose). We found the usual dog tracks, with mink, fox and deer, as well as interviewing park users. We interviewed firemen who were assessing old mine shafts, who told the class that large feral cats were present in the area. Another interviewee informed the children of a chance encounter with a wildcat whilst in the Middle East.

Exotic pets and potential releases / escapes were looked at, and again the children researched sightings in Britain. Potential tigers in Yorkshire, and leopards in Lancashire, causing a bit of a stir! As in the Bolton project, we also had to consider a risk assessment for a school trip searching for big cats!

The final artwork was exhibited in the school reception for several weeks, then at Manchester Museum, and then installed at Salford Museum and Art Gallery for a month in June 2007. As with the Bolton Museum project, the final work referenced scientific research and museum display techniques in order to present as much of the children's efforts as possible.

The project went really well, and - in fact - was so successful that the theme of native, and non-native, is now to be taught in Science at the school a year on.

The potential for using 'alien big cat' sightings within schoolwork is enormous, as it is a remarkable subject, and with a bit of imagination can be linked to so many subject areas. Within the two projects I worked on, we covered Art, Literacy, Science and Citizenship. On both projects, the children were completely engaged with the idea, and indeed the whole school staff enjoyed the subject. I was also lucky in that the people I worked with, and pupils of both classes, were very enthusiastic, and great company.

As to whether there were any big cats out there - museum and wildlife staff mostly had an open mind. Though they had not seen a body (and museums get a fair amount of RTA wildlife) they were of the opinion that it would be possible for these cats to survive relatively unseen. The North West has ample wildlife, and large areas of countryside in which to live. Some of it very near to schools, galleries and museums...

Every Village Should Have One
by John Beart

In 2005, an article with the above title appeared in the *Daily Telegraph* written by a journalist living just north of Cirencester. It discussed, rather sarcastically, reports of a big cat in the valley where he lives, and the fact that there are an increasing number of sightings, suggesting that they were imagined, and that the whole thing is a rural myth.

About three and a half years ago, returning from visiting a friend who lives in The Forest of Dean, I was travelling on a narrow back road through the woods, when a big cat emerged from the bracken, and crossed the road about 50 m in front of the car. It was July, late afternoon, very bright, excellent visibility, with no other car in sight, and I was going up a long incline so it was about level with the windscreen, giving me a perfect view. As I passed the place it had crossed, I saw it from behind climbing up a bank into the trees. It had jet-black gleaming fur, the head and body shape of a cat, and the unmistakeable long looping tail, and it moved like a cat. It unmistakably was a member of the leopard family, with absolutely no possibility of mistake. I cannot describe my shock. I was thunderstruck, astonished and bewildered, as I had absolutely no idea there could be anything like it moving around in Gloucestershire. Once home, my wits returned, and I rang my friend to tell him what I had seen, and he told me that there had been a number of sightings reported in the area, and suggested I tell the local paper and a man called Danny, who was keeping a log of sightings in the area. This I did, and was surprised to learn just how many sightings there had been. As time passed, I tucked the memory away as a once in a lifetime experience, though I confess that since then, I have kept a wary eye when out walking the dog.

Now what is the point of me telling you this after several years? Well, read on.

I often take my dog walking in the fields around Witcombe, fairly early in the morning, or in the late evening. There was one very big field where I could let him run, and I would sometimes sit on my shooting stick and look at the newspaper, looking up occasionally to keep an eye on him. One of his favourite tricks was to pop through the hedge near me, then run along the other side, and pop back out again at a footpath about 150/200 m further along. One morning, about 21 months ago he was playing this game, when I looked up and he had disappeared (he was right behind me, sniffing in the hedge, I later discovered) so I looked towards the footpath expecting him to appear, and there running across the field was what looked like a big cat. I think it saw me stand up, as it speeded up, disappearing from view through the opposite hedge. It was large, jet-black, and looked - and moved - much like the previous one I had seen, but it was much further away, and going faster.

Therefore, although I was almost sure that it was a big cat, because of the distance there remained just a small uncertainty. I went home almost as shell-shocked as before. I thought perhaps I should mention it to a neighbour who lives next to that field, and unexpectedly he was not surprised and told me he had previously found really big paw prints in the snow, and other evidence of a big cat on his property. I also, later, mentioned it to the vet, and he said there had been reports of one in the Cranham area, so I was more certain it was a big cat that I had seen. However, near Witcombe, very hard to believe! As the chances of seeing it again were so remote, I thought I'd never be 100% certain.

Well is that it?

No. At the beginning of June this year - late one evening - with still reasonably good light, I walked the dog up towards the fields, glancing warily into them as I always do now, and there - through a gap in the hedge - up on a bank about 15 m away, was a big cat moving quietly in the opposite direction through the field parallel to the hedge. It must have seen me, but showed no reaction. Also, there were two horses in that field further over and, surprisingly, they showed no signs of anxiety, though it is possible they were looking away. I thought about that later, and decided that they possibly had seen it before and so, perhaps, neither horse nor cat regarded the other as a threat. Once again, I was utterly amazed, and shattered, at what I had seen - so close too - especially as it was fairly near to my house, and several others.

Thinking about it later, I became rather concerned that I now had three similar sightings when most people have not had one. If someone had told me that they had seen a big cat on three occasions, frankly, I would have had very serious doubts about them, and the reliability of the sightings. I suppose though, I do go walking in the right areas at the right times. Anyway, I reported this last sighting to the police, who took details and said they would pass it on to DEFRA. On reflection, I thought that surely there must be other people who have seen something. Then a few weeks later, I bumped into a man that I know. He lives up towards Birdlip, looking out over the area of my sightings, and so without any warning, I asked him if he had ever seen signs of a big cat. Without any hesitation, he said *"yes"*, and I know him to be a serious and honest person. He then said that he was walking up through the woods above Witcombe when he got an odd feeling that he was being watched. He glanced into the trees, and there was a big cat laying down between the trees, about 4/5 m from the path, watching him. He said it never made any movement; just its eyes followed him as he continued walking. I made no mention to him about any of my sightings, but I just asked him the following questions. When? –*"About two or three months ago"*, (about the time of my sighting). What colour was it? –*"Jet black and it was really beautiful"*. What size was it? – *"Big"*. Did you tell anyone? -*"No, but I don't walk up there now without a stick"*.

Now you can dismiss all this entirely or be very, very sceptical - I can quite understand. I have thought long and hard before writing this, as I have no wish to cause anxiety or alarm, but I thought most sensible people would want to know, especially those that walk up in the hills. Should we be alarmed? Well, I think that the cat (perhaps there are more than one), has had more sightings of us than we have had of them.

However, they appear to avoid human confrontation, go quietly about their business, and so seem unlikely to be aggressive unless provoked, feeling threatened, or perhaps having cubs. I know they ignore horses, but I am not so confident about dogs, or little people. There are plenty of deer and wildlife about, so they should be well fed, but they are large, powerful and potentially dangerous creatures. Assuming they have a range of 5 or 6 miles from "home", that gives a huge area of up to 100 square miles, so bumping in to one seems unlikely. However if do you see one, I suggest you ignore it, do not stare at it. Just walk on, or let it walk on, and do not make aggressive moves unless you must, but that's only my opinion.

Why has it been seen several times in the Witcombe area? I have no idea, but I suspect that, like foxes, they patrol well-established trails and we periodically get a visit or maybe, like the one in the forest, they come down for a drink.

Well there it is, so it is not just me, and I suspect there may be others who keep quiet to avoid being scoffed at. I do not intend to give my name, as I have no wish to be plagued with questions for details, or by the media. However, if you have had any clear sightings or useful information, then pass it on through Tore or Caroline, and I will compile a record. In addition, I believe there are websites. I suggest you search under 'big cat'. I looked three years ago after my first sighting, and there was a lot of information then, but I have not looked since. There have been almost 3000-reported sightings in the last year, and probably as many - or more - unreported. If only 1% is correct, that makes between 30 and 60 big cats, and that is cause for serious thought!

John Beart is a retired engineer who lives in Great Witcombe, near Cheltenham, Glos.
This article was first published in Witcombe Village News
For *British Big Cats Yearbook, 2008*

"If There Were Big Cats, I'd see them!"

by Rick Minter

Rick Minter edits ECOS magazine (www.banc.org.uk) and runs the environmental consultancy Nature with Attitude. He helps investigate big cats in Gloucestershire, on which this review of 2007 is based.

Working the grapevine

As ever, 2007 was a year of routine sightings of big cats in Gloucestershire, several of which tallied, with remarkable consistency, in terms of timing and location. As soon as it goes quiet, another email or phone call crops up to reveal the story from another witness, and keep everyone guessing as to what is going on. I have not been in this game long compared to many big cat specialists in Britain, but already I am learning not to expect a breakthrough any moment, such as startling emphatic evidence, close-up footage of a big cat smiling to camera, or a carcass from a road kill – these are the decisive factors that all the investigators yearn for. Not that we have not had some tantalising moments in the County this year...

Set out below are just a few of my thoughts, prompted by the year's events and incidents in Gloucestershire. They are unremarkable personal reflections, and many will strike a chord with the experience and insights of other big cat researchers and enthusiasts. The common thread is the human reaction - the nature of people's responses to the big cat phenomenon. This is a constant fascination to me, and no doubt to many other researchers. It is something I find as intriguing as the physical evidence of the cats, and their field signs. It is the witnesses, and the grapevine, which spreads the word about sightings, and our relationships with these networks of people, which largely feed us the information we have to go on. In the absence of proper resources for systematic filming and rigorous field investigations, we will continue to need to operate like this, so we had better keep plugging away at it, and we might as well enjoy it.

"These blokes are talking about big cats – it's ridiculous"
Ecologist hearing me converse on big cats during 2007.

I am used to being shot down about big cats by some of the ecologists, and wildlife conservationists, that I meet professionally. I quite understand their view. They have their subject utterly sussed, and do not need the likes of me, a touchy-feely type, telling them that there is more to Britain's ecology than they reckon. To them, the prospect of leopards, pumas and lynx, or some other exotic felines, somehow living free in the British landscape, appears fanciful and nonsensical. It questions their authority, it messes the picture of wildlife they know and study, and worse, it takes them right out of their comfort zone.

I find the sceptics, like these folk, come in two camps. Most of them are decent and polite, and demand the evidence (and why not? They are scientists who deal in tangible, certain things). Others (far fewer) are hostile, and just counter and dismiss me. Their understanding and perception of what is in our countryside will not be challenged or complicated, so awkward sorts like me get short shrift. I have come to terms with both types of response, and I am pleased that there are some 'aggressive sceptics' out there. They should remind us of how flimsy to date the ultimate forensic evidence is, despite the relentless waves of reports and sightings, which come our way.

My favourite 'assertive' reaction during the past year was the chap who claimed there simply could not be big cats in our landscape: as a field ecologist, he was out all the time, in deep, remote countryside: *"If there were big cats out there – I'd see them!"* he proclaimed. However, as big cat researchers know, some farmers, foresters, gamekeepers and deerstalkers (people who are out there, in deep countryside all the time) have seen them, heard them, or flashed a spotlight on their eyes, while others have not. There is nothing certain about it – it just happens or it does not, or it has not happened yet. Melting into the scene is precisely what these cats do. A lynx tracker in Bavaria, whom I met this year, routinely tagged lynx for telemetry, after catching them in humane leg traps. However, in the wild, right in the heart of its territory, this chap, so finely attuned to the animal, had seen lynx just three times in eight years.

The same doubting ecologist heard me swapping big cat stories with a mutual friend this year at a professional gathering, and he could not hold back his horror: *"These blokes are talking about big cats – it's ridiculous"*. We were frowned upon, and made to feel like naughty boys at the back of the classroom. I want to stress that I have no hard feelings towards this chap - I find him good company, and if he feels that talk of big cats in Britain is "ridiculous", no doubt, others do too, and it is just as well I get this reality check.

I have had an ecologist friend change his mind on big cats this year. He'd previously taken a cautious 'Where's the proof?' approach, but to his credit, he's accepted there must be something in all the sightings which fill his local paper, *The Forester*, in the Forest of Dean. There are two other factors, which have helped sway it for him: the consistency of the reports, and the movement and jizz of the animals that people describe – unmistakably a cat's movement, as he notes. It's usually an actual sighting which triggers a change in view amongst sceptics, so fair play to my friend Martin, and I only hope he's rewarded with a sighting, although I'm not sure he's that bothered!

"Wow - I'm believing this – I think!"
Radio Gloucestershire presenter, interviewing me in May 2007

It is heartening to be accepted and wanted by your local media. I guess a resident "expert" (the media's term, not mine) is a useful resource for the media, to give them a handle on a mystery subject, which arouses people's interest. And we have other respected investigators in the County too, in Frank Tunbridge and Danny Nineham, who give quotes and interviews to papers, TV and radio.

I do not get any sense of the local media (the Gloucester Citizen Newspaper and BBC Radio Gloucester) treating me as an oddball or eccentric, and if I did I would be wary of dealing with them. In my view, if the media are keen, and want to report a topic 'straight', then working with them is a valuable way of influencing people, and of getting feedback and information – and, crucially, more sightings. Seeing it like this, any slight risks cannot compare to the advantages.

Radio Gloucestershire have interviewed me five times in the past two years, including a 40-minute live stint in May this year. In this 40-minute session the interviewer excelled – he tossed up all the right questions, giving me a perfect platform: How do these creatures live and exist? How did they get here? Do you speak to government reps about it? What's the best evidence we have in the County?" As he cut to a brief traffic report, he boldly declared: *"Ok, more in a minute on big cats folks – hold on to your seats: Wow - I'm believing this – I think!"* The obvious then dawned on me: for many, if not most listeners, the session they had just heard would have seemed remarkable – a local man daring to describe in detail how large exotic cats live freely out there in the local countryside. Really? Seriously? In addition, as the traffic report unfolded, I thought more about how people might react – most people listening would hold a strong opinion one way or the other. It is not the sort of subject you can easily ignore or be half hearted about. People out there would think either I was a complete nutter, or they would give me some respect. They might be scoffing at stories they felt I was concocting or exaggerating, or they might be relishing such straight talking on a shadowy subject. I just hoped there were more of the latter.

For the second half of the interview I was invited to respond to several people's phone-ins of their sightings: a taxi driver seeing one jump over a hedge; a lady farmer having her Jacobs sheep devoured, then seeing a "bobtail lynx" through her spotlight; a lad from the Forest of Dean watching a big black cat saunter through a field of alarmed sheep (why didn't it stalk them?) when out with his air rifle. If I was indeed a nutter, then I was not alone amongst the good people of Gloucestershire. As I left the studios, I was told they were pleased with the whole session - the phone line had had a queue of callers, and the station had detected lots of interest.

As good for the ratings as big cats might be, I think we must give credit to the likes of *Radio Gloucestershire*, and to any radio and TV stations that cover the subject – treating an unorthodox topic like big cats must be far from straightforward for these people. The more we can give them a sense of authority and confidence, hopefully the more at ease they will be in handling the topic and being willing to pursue it.

"Isn't it like knowing a good beach – shouldn't you keep all this to yourself?"
Farmer in the Cotswolds, Sept 07

Since November 2006, I have given five talks on big cats to different audiences around Gloucestershire. I do a double act with Frank Tunbridge, who has been following up big cat reports for over 20 years in the county, and who gets plenty of sightings passed to him at his weekend car boot sales. Frank provides the perspective of a seasoned tracker, and he is able to sprinkle in plenty of colourful anecdotes. He can illustrate any type of sighting you could imagine, from children having their play interrupted by a panther's presence; milkmen startled during their delivery round; a chap charging off in fright when a panther jumped out of a tree; and drivers having to stop at the dead of night because a big cat just would not budge from the roadside. Frank has an endless supply of such close encounters - his report of temporarily catching a woman's dog in a baited cage, falsely branded by the local paper as the device of "evil badger baiters", soon has everyone chuckling. In the 90-minute sessions, I add some analysis and a national overview, and I show some photos. Frank also does sound effects – few can better his vocalization of a leopard's cough, and a puma's call to cubs.

Giving these talks has become an important anchor for us. It keeps us linked to people and places around the county where we can gauge what has been happening, get more reports, and answer queries as best we can. The talks are a vital way of gaining people's confidence and building a constituency of contacts – many of our reports now come through people who have attended the talks, and who pass on sightings. Some people have even had sightings since hearing us. One woman witnessed two big black ones crossing a road together in early November 2007, and had to slow her car as she approached. Wasn't she surprised after such an awesome encounter? *"Not really, after hearing your talk!"* A week later, a trader at one of Frank's car boot sales reported a sighting of one big black one crossing the road within 200 m of the same spot, in the same week.

I often wonder at people's motives for attending the talks. Thankfully, few seem to think we are mad, and some have their own sightings to describe and discuss. It is rewarding having all manner of people present: we get a genuine mix from all sorts of backgrounds and ages. Near the front, it is common to have children present – some have had sightings themselves, or know a friend or family member with a believable report. The presence of children is welcome, and it concentrates the mind. It reinforces the need to be measured. Of course, we have to make talks informative and entertaining, but the message, I believe, must show respect towards the animals in question, it must promote cautious and responsible behaviour in any situation where big cats might be present, and it must avoid anything, which could be taken as scare mongering.

I am used to public speaking and being involved with gatherings of people, but there is something different about big cat talks. There is a special kind of response from the participants. From their body language and expressions, many people seem absorbed by it all - if not spellbound. As passable as Frank and I are at giving talks, I put this down to the subject, not us! We habitually over-run the allotted time, but nearly everyone stays late, and remains for the final questions and witness reports. Even afterwards, while we are packing up, many people linger on; wanting to chat amongst themselves, and posing further queries. It can be tricky to get some people to leave – it is as if they want to savour all the time amongst the kindred spirits they have met. It all makes it flattering to be involved.

The comments and questions at the talks are as useful to Frank and I, as they are to the audience. They illustrate what is on people's minds, and what they have and have not fathomed out for themselves. There are usually plenty of gems too. Below are some, which stick out.

Handling the hecklers

Our first talk was to a large group of farmers. Most were interested and attentive, and some had good sightings to report at the end. But, during the session, it was evident that a few were scoffing at our comments. They could not hold back their disbelief, and their laughter punctuated the talk, creating some awkward moments. At the conclusion to the evening, they were confronted by the group's Chairman – a respected Cotswold farmer - well known in farming and wildlife circles. As a leader of their peer-group, he dropped the following bombshell - a direct bull's eye on the hecklers:

"Anyone not believing Mr Minter and Mr Tunbridge is welcome to come on my land in the Cotswolds, where my deer stalker will show you around, hopefully from a safe distance - big cats have been seen on my farm for the last nine years."

Keeping it to yourself?

Some people would rather not face up to big cats being around. They feel it is something we should keep to ourselves, not broadcast, and not venture an opinion on. I understand this view entirely. If a topic is so unofficial, and surrounded by so much speculation, why 'come out' and pronounce on it, and remind people about all the issues involved, none of which can be dealt with anyway? One farmer, who had seen a big cat, and knew plenty of others who had, put it perfectly after a talk in Cirencester in September: *"Isn't it like knowing a good beach – shouldn't you keep all this to yourself?"*

My reply was that I was conveying what I felt needed to be known, and I'd rather a (hopefully) responsible message came from me, rather than less measured information from anyone who might have a different agenda on big cats. The farmer, and the audience, were very decent in response - they seemed to accept my rationale, although I am always happy for it to be challenged.

Counselling the witnesses

When big cat investigators interview witnesses, it is rarely just an exercise in note taking. The investigator often interprets what might have happened, so the witness can put the extraordinary experience they have just had in some sort of context. For a witness, it can be comforting to speak to somebody who will believe and understand him, or her, and take time to listen and respond. Indeed, many investigators recognise that their listening role amounts to counselling, when taking witness reports. A vivid example occurred at a talk I did with Frank in February 2007. A woman had travelled from neighbouring Herefordshire to be present, and amongst the questions, she explained why she'd come along.

She had had a remarkable and shocking encounter 10 years ago, when walking her dog in a forest ride:

"My dog froze. Then I realised why - although still on the lead that I was holding, a panther at the edge of the wood, was stalking him, close by. I could not believe it and was terrified. Eventually I think it recognised there was a human about too, and pulled out of the attack. It bolted away along the ride at dramatic speed. It was 10 years ago but the moment still haunts me now."

Being at the talk, hearing of other people's encounters, and getting feedback was clearly a help to her. During the packing up at the end, she was able to spend a good 10 minutes discussing it further with Christopher Johnston – an experienced investigator who'd come along to the talk too, and who knew some similar situations faced by witnesses and dog walkers. Chris felt that she'd left the event with the whole troubling experience now exorcised.

Getting out more

The great outdoors is where I yearn to be, but I do not get out nearly enough. I've absolutely no excuse - I live amongst open fields, have a network of footpaths all around, and stunning Gloucestershire landscapes and wildlife-rich nature reserves are just minutes away by car. This grumble will sound familiar to many big cat investigators (and indeed people of all persuasions) who, like me, are too often slumped over the keyboard. Big Cats in Britain coordinator Mark Fraser rightly nags people to get out more - the evidence will not be found until we do, he reminds us.

Thankfully, during the year I woke up to the use of stealth cameras (or remote cameras). I now have a motive for a regular wander up to the local wood, where, luckily, 'big black ones' have been seen. Even more luckily, for my purposes of installing a camera, the local wood is nicely tucked away, and hardly gets any walkers, poachers, or whatever passing by (I am otherwise very pro people using and visiting the countryside - honest). The farmer and farm manager enjoy the novelty of having the remote camera snapping away on their land. They advised where it should go, in recently planted saplings providing a belt of extra cover next to the wood. The area is criss-crossed with tracks (mainly fox) and the camera is tucked away, close to where the farm manager and his daughter spotted a big black cat in early September, the same evening another sighting of a similar description occurred a mile away, an hour earlier.

I previously had a stealth camera set up on the scruffy urban fringe of Gloucester, where Frank knew from sightings and sounds that there was a jungle cat present, and where Frank had once seen a big black cat swish through the swampy vegetation, and heard a leopard's cough when it went under the bridge where he stood. It is a great site for wildlife, with secondary woodland, an extensive wet meadow, riverside vegetation, and lots of rubble from old pipes and demolition work, heaped into rock formations. The range of cover, plus the abundance of rabbits and deer (and perhaps some of the birds) makes it nice territory for cats of any size. It would be perfect for regular use of a camera, but alas, it is bandit country: teenagers hang around in the evening and people do unlicensed fishing for elvers along the Severn. It is no surprise my first camera was vandalised - Frank was outraged to find it in pieces on one visit.

For a while, at least, my camera will stay in the local wood. It is conveniently close, and the farmer is happy - that makes for a good combination. He, our families, and I enjoy just seeing the photos of fox and the deer, and the anticipation of what will appear next time. The first pictures of Mr Fox showed him extra cautious - he looked on edge in every shot, perhaps getting used to the camera's smell and the flash. Later shots have him at ease, and in one, he even appears to be casually smiling to camera - perhaps saying thanks for all the sardine bait he pinches. The deer now pass by as well - we have some half-decent pics of roe deer and (a surprise to the farmer) a muntjac. Not surprisingly, we get squirrels, magpies and pheasants too. Ok, these creatures are not the target, but it is gratifying to get pretty pictures of the local wildlife, and who knows - their tracks may one day draw in an alpha predator!

For a sporting chance of snapping any passing cat, I ought to be adding a couple more cameras around the wood. As well as £120 for each bit of kit, this will mean more expense on batteries, more lavender oil drizzled on the surrounding trees, and more tins of sardines - bait that is instantly snaffled by the thieving fox. But this all seems a small price to pay. For softies like me, who are hapless trackers, and reluctant to camp out, the remote camera is affordable, and pretty much hassle free. I am now ready for 'the beast'...

Meanwhile, in the Stroud valleys...

As I write in early January, three sightings come into together, all from the same square mile, over the same week, southwest of Nailsworth in the glorious Stroud valleys. They all seem pukka accounts. On one, along with the sighting, there were dogs spooked ("frozen into immobility") and on another, a husband and wife watched the big black cat for 15 minutes, walking and sitting by a hedgerow. Crucially, they have come from two different sources, independent of each other. What do my scientist pals think of that?

Territory or Range

by Christopher Johnston

A territory is a geographical area that an animal will actively defend to protect the resources within it, like food, mates and safe resting places. A range is a large geographical area that is too big to be classed as a territory, and would require too much energy to actively defend, and would leave less time to look for a suitable mate.

British big cats are difficult to study in a scientific way, tracking them, radio collaring, and collecting data is just not possible at this present time. With remote cameras becoming cheaper, and more easily available, more researchers are using them, improving the chance we have of obtaining some good results, but the placement of these cameras is the key and, in itself, is a challenging task.

Cameras can be placed at a recent sighting, but it's not known if the cat will stay in the area, will return in a few weeks time or may never be seen there again. I have researched sightings in many parts of the country, and in some areas it might be possible to say that a big cat holds a territory, for example Bodmin Moor and Exmoor. Big cats have been seen regularly on the two moors over many years, and evidence found, but is this one cat on each moor or is it the one cat moving from one moor to the other, or a small number of cats moving from one moor to the other?

It is important to consider that when an animal is placed in a totally new environment its behaviour must change to adapt to its new surroundings - if it does not adapt then it will not survive. Over time other changes may take place. This could include changes in coat colour - a possible consequence of inbreeding - and if this mutation is beneficial it may become the norm.

Only as an example we will look at the behaviour of a leopard in its natural habitat. We can then compare this to life here. A leopard's territory can be from 5 to 30 sq miles - the number of leopards in the area, and the density of prey, are factors in determining the territory size. This diagram shows a map of a leopard's territory as it might be in Africa. The black line is the male's boundary and the grey is the female's.

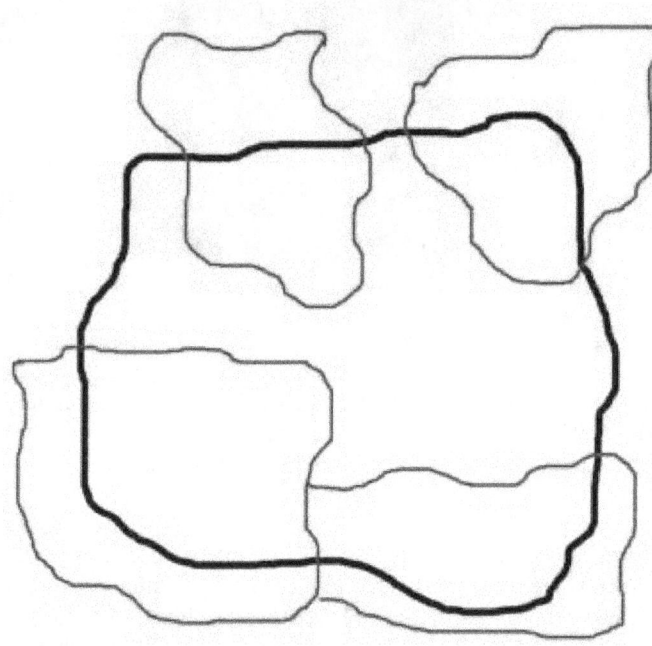

This map is only possible if there are the numbers of leopards to complete it - a map of territories in Britain could look very different. Males have larger territories than females, and up to four females may live inside the territory of the male. The female's territory may also overlap into another male's territory, as this gives the female a choice of who will father her young. The desire to mate and reproduce drives all species, passing your genes on to the next generation is important. A leopard living on Bodmin Moor might not have a suitable mate in the area, and then might have to travel to Exmoor, or even further, to find one.

If this was the case, the territory would then become a range. This could have an impact on their behaviour; less aggression would be required along with not actively defending a territory. Leopards and other big cats are not lazy, but will conserve energy wisely; travelling greater distances to find a mate would require more energy, but this is available to them in our countryside and could allow them to cover a large geographical area. Sightings would then be over a greater area, evidence would be more difficult to find, and photographing them with a remote camera even less likely. This behaviour could also suggest that there are more big cats living in our countryside than there really are, with cats being reported in one county, but then being reported in another as they move through their range.

Communication would still be important as they are solitary, and would want to avoid one another unless looking for a mate. Making a scrape, spraying vegetation and leaving a scat as they move around would allow one leopard to let another know that, "I am in the area today", and avoid any confrontation. Again, only as an example, there could be only five leopards living in the Southwest of England all sharing a large range, moving from Dartmoor to Bodmin to Exmoor, and communicating with each other as they go.

It is possible to observe this exact behaviour in a big cat species living in its natural habitat, e.g. snow leopard. I am not saying that there are snow leopards in Britain, but leopards and puma living here may have adapted their behaviour, and show similarities to the snow leopard. Snow leopards have a large range, and do not actively defend a territory as such, they communicate as they move around to let other snow leopards know they are in the area. They are less aggressive than many other big cat species. British big cats even have the benefit over the snow leopard of more readily available food, maybe even reducing aggression, and competition, even more.

I have seen a puma on Exmoor - in February 2007. The area it was in provided a safe haven, with warmth and food at this time of year. In the summer, this remote and quiet area becomes busy with holiday makers, and walkers, and in this season would not be a safe place for a puma. Sightings have been recorded at this location over a number of years, and all are during the winter months. Having a large range would allow this puma to travel to an area more suitable for the time of year.

A range opens up more possibilities to them - they can move to suitable places as the seasons change, find prey whatever time of year it may be, travel long distances to find a mate, and not have to worry about defending a territory. Less chance of having aggressive confrontations may improve their longevity.

There is no evidence to suggest that what I have written has any truth, and there are many more scenarios regarding territory size. Hopefully, one day, it will be possible to answer this question in greater detail and with more accuracy.

Black Panther Killed On Bypass
by Frank Tunbridge

Frank Tunbridge has been investigating big cats in Gloucestershire for 25 years. He receives witness reports on a routine basis, and regularly advises witnesses and follows up sightings. He has had several sightings himself, and has collected and photographed field signs of big cats.

"Black Panther Killed On Bypass"

How often do these headlines present themselves? Never! But it has happened on several occasions throughout the UK. It amounts to definite proof that these creatures are living wild in rural, and urban areas, of Great Britain. What happens to the corpse is another story.

Police forces, DEFRA, highway agencies, local authorities, and government departments all become involved where the issue of public safety and awareness is concerned. It is a mystery and will continue to be so, until a *number* of captives or corpses are presented for scientific analysis, and media coverage. Ask any citizen of the UK if they believe that we have wild boar living free in our countryside and most will confirm this fact, having now seen the boar on television programmes such as *Autumn Watch*, and read about their existence in local and national newspapers. Some people have even encountered them personally.

But mention big cats living wild and thriving amongst us then the answer is generally - 'Yeah! Right!' But, and I make a statement now which will surprise most people reading this article, out of a cross section of members of the public living in rural areas questioned whether they have either sighted, or know someone who has sighted a big cat, the figures come up as 15 out of 60. That is a quarter of the number quizzed on this subject, who gave a definite answer. The other 45 must belong to the 'yeah right' group. If any of these sceptics were ever to encounter one of these elusive and stealthy animals, their comment would change to '*yeah,* you *were* right' instead. Anyway enough of this, let's move on to other evidence.

Deer Kills

If these big cats are in any way derived from leopards and pumas, to name just two species released into our countryside over 30 years ago (I believe most are now breeding true to type - I will move on to that later) antelope and deer would be their preferred prey. We have no native antelopes in this country, so deer, of which we have plenty, are the natural prey for these animals.

When a deer kill is encountered, many times it has all the hallmarks of a large predatory cat kill: asphyxiation, or broken neck, deep puncture wounds to the head and neck area. The corpse has often been dragged into cover, bones licked clean, large claw marks on the flesh and a large amount of meat consumed in one session. It is very unlikely any other native British predator or domestic dog could achieve these impacts. Also, remains of deer kills have been reported lodged in lower branches, and forks, of trees. Indeed not long ago, an unnamed deer poacher rang me to confirm that he had found blood and deer hair where a deer carcass has been carried up into a large pollarded willow tree, located on a large country estate that he poached regularly. The same chap also mentioned that when he was skinning a fallow deer taken from the same area, very long claw marks were found on the haunches of the deer. This guy had nothing to prove - like many others who report sightings and evidence to me, including farmers, gamekeepers and forestry workers. These frequent occurrences are accepted as just part of their working agenda, and are often heralded as a possible threat to their stock, rather than a rare sight in our green and pleasant land.

To add more credence to the view that deer, especially fallow, sika, roe, and the snack-sized muntjac, are these cats' preferred prey, I point to the fact that on many occasions throughout the UK, deer sighted running across the road have been hotly pursued by a very big cat. Gloucestershire, Devon, Dorset and Wiltshire, to name a few local counties, are where such reports have come from.

The more abundant a prey species becomes, so becomes the predator that preys on it. I have investigated several reports of deer kills over the years, and - apart from a few road kills scavenged by foxes, crows and some obvious dog attacks - a large proportion of these deer were the victim of a large cat-type animal, with all, or some, of the hallmarks mentioned earlier. The old survival adage applies anywhere in the world: where there's something to eat, there's something to eat it. This dictum applies right throughout the food chain, to the larger predators at the top. So do we now have a top predator in the UK? The big cat - sometimes elusive, sometimes bold, but always silent and stealthy - seen singly, in pairs, with cubs, and in rural and urban environments, taking a wide range of prey from small mammals to large deer? These cats are often seen, but - to date - have seldom been caught by camera or video footage, and then only at a distance.

The origins and the species

Another controversy concerning these anomalous cats is the origin of their presence in this country, and also what species they are. On the first issue I believe that the 1976 Dangerous Wild Animals Act just added a great deal of fuel to the fire as there were already a number of big cats established in our countryside by then. I believe that by dint of introgressive hybridisation, a naturalised** type of big cat now exists in our countryside.

The uniformity in the sightings, that are received weekly, seems to confirm this. The descriptions of the black type of cat sighted are so similar, across a large area covering several counties, that this is more than just coincidence. The usual description reported hardly ever changes. Here is a typical, and in my view, apt description: *'it was like a stretched Labrador, small head, moving like a cat with a very long tail, jet black and shiny, like an oil slick crossing the road.'*

Most sightings of these very big cats occur at dawn or dusk, although they are occasionally spotted during daylight hours, but - as is usual with many predatory animals - most activity is confined to being crepuscular and nocturnal. The obvious explanation is that their night vision is far superior to that of their prey. Also, mankind is much less likely to be moving about at these hours, leaving them undisturbed to go about their business of hunting, mating, territory marking, and the like.

I have followed many sightings of big cats over the years, and in almost all instances, the encounter has been unexpected. They see us before we see them. *'A hundred ears hear a single footfall in the forest*' it is said.

How big is a big cat?

From various sighting reports from all over the UK, these cats vary enormously in size. However, from my own research I have found the larger type i.e. (Labrador, German shepherd dog size and above) tend to live in more rural and remote areas such as Exmoor, and parts of Devon and Wales, where large red deer and ponies (especially foals) often become the prey species for them. Smaller big cats, between fox and Labrador size, often frequent town and urban areas where the smaller pickings are at hand. In the book *Field Sports of India 1800 - 1947*, author Dunbar Brander, a renowned hunter and naturalist, states that a leopard's size is dictated by its diet, and if a high percentage of these big cats are derived from leopards originally, the above holds sway. Lynx are often reported seen in quarries and other elevated areas. They seem to have a penchant for such places; they also are very secretive and elusive, but a predator through and through.

(** Appertaining to any wild animals species that has become established in an area to which it is not normally native).

Territory, size and behaviour

Every animal has a territory or home range, from a mouse to a moose. Cats are no exception. Their range is dependant on factors like availability of food, water holes or drinking places, and the distribution of other big cats in the adjacent area. Mating also plays a vital part in their distribution, and their home ranges will often overlap, especially when they are ready to mate. Overlapping of territories will affect behaviour like scenting, scraping, and vocalisation, as well as other behaviour.

These big cats must be breeding in the UK mainland, as both adults and cubs have been spotted on many occasions. Plus, two adult cats seen together, and a mother with cubs, are frequently reported.

The kind of food a species eats will eventually affect its anatomy, and also its method of hunting, if necessary. Habitat, and the lack of prey, will also determine behaviour as has been found with feral domestic cats. Far from being the solitary hunter like the Scottish wild cat, feral domestics often tend to live in well organised communities similar to lion prides.

Big cats of Border collie and Labrador size are often reported seen in my area of the country. These are generally black in colour, and sometimes seen in pairs. Apart from females with growing juveniles, I believe this behaviour could, in some instances, be a new trait of two adults co-operating in a hunt that has developed over the decades living on our island. Such a trait of working together allows the bringing down of large prey like roe or fallow deer.

Why black? Why not?

Everything goes with black as they say in the fashion world. Apart from the recessive gene theory, black is ideal pelage for a stealth hunter. Just one step back behind a screen of vegetation, and it blends into the shadows. It is also ideal in urban areas where concrete and tarmac abounds. By the way, have you ever noticed that the majority of domestic cats seen in towns and cities are black, or black with some white markings?

Many of these big cats opt to live near the outskirts of towns, and frequent urban areas, with many sightings being reported from building sites, industrial parks, and pockets of urban fringe waste land etc. I found evidence last year of a big cat type animal residing in a derelict site within an industrial estate just off a busy bypass. The site was overgrown with brambles, and other vegetation, and supported a large rabbit population. The animal was behaving like an urban fox: stealthy, opportunistic, taking advantage of what food was available, such as rabbits, voles, mice, rats, feral pigeons, and the occasional discarded takeaway. Reports like this are quite common, and one sighting reported to me was of a big cat seen eating some leftover fish and chips at the side of the road!

One other aspect of these cats' behaviour has often come to my attention. Many of the reports I receive over the course of a year point to the fact that they have been seen during, or just after, a violent electric storm. Also I have found that these creatures often follow the course of overhead power lines for some reason.

Public safety issues

At the moment nothing serious has happened to any person(s) within the UK; but if a big cat were to be injured, and unable to catch its normal prey, more livestock, pets and then perhaps people, especially the very young and very old, could become part of their diet. So far, only a handful of people have been subjected to a minor attack by a big cat type animal in the British Isles.

Automatic trip cameras

Up to now, photographic evidence of big cats living wild in the UK has been limited to a fairly motley selection of images, with only a few providing a description of an animal that can be identified as nothing else other than one of these elusive big cats. Much footage of these creatures conveys out-of-scale images, or vague images in the distance. Automatic trip cameras are one way of obtaining more conclusive evidence of these animals' existence. The person setting these up ideally needs an intimate knowledge of the animals' movements around their home ranges, to be able to allocate them in the right area, where a big cat is likely to pass by. This is similar to the efforts to obtain footage of the other elusive big cat, the snow leopard. This venture was highly successful, providing excellent images of these animals, yet positioning the cameras took many months by professionals solely dedicated to the task.

So, big cats in the UK are here to stay existing alongside returned species like wild boar, or recently naturalised exotics like ring necked parakeets and the like. I predict that within the next year or two, so much overwhelming evidence of these elusive big cats will be presented to the media, via video footage and physical specimens, that the present 'mystery' hanging over these creatures' existence will be solved.

Bringing Big Cats into the Work Place
by Chris Hall (Teesside)

One of the benefits of the Internet today is the wealth of accessible information available to view on any subject; perhaps more importantly, many websites have facilities to either leave comments or send reports, which I think is one of the reasons why there are so many big cat sightings coming to light these days. One only has to study 'Big Cats in Britain', to see the wealth of new sightings now available.

However, sometimes having the right job helps also. I work in the retail trade at a large grocery superstore situated on the western edge of Stockton on Tees. I work on the checkouts, and also do the home shopping deliveries in a van; both these roles give plenty of opportunities to bring the subject of big cats into the conversation. The value of these, hitherto unreported sightings, is immeasurable. So far, in my shop alone, I have had several customers, and three members of staff, talk to me about their experiences.

My store's catchment area covers many rural areas, and several towns and villages in this part of South East Durham and western Teesside. The home shopping vans, however, operate over a much greater area, reaching as far south as the northern edge of the North Yorkshire Moors, and almost as far as Whitby. Whilst all these areas have a history of big cat sightings, the purpose of this article is not to list these, but rather to illustrate how easy it is to get people talking about big cats in the workplace.

The staff bit was relatively easy. All I did was to produce the yearbook, or one of my other cat books, whilst on my tea break, and read it. Eventually someone shows an interest, and you get talking. Sooner or later you hear the magic words '*I saw one once*' or '*my dad saw one on the way to work*', and there you are, a sighting! Jot some details down, and the cat's in the bag so to speak!

The customers are a bit different in that you have to exhibit discretion. However, by selecting the right sort of customer, you can usually introduce the cats. I look at the type of person, and what they buy. For instance, someone who turns up in jodhpurs, or a Barbour jacket, is likely to be an agricultural worker, or an equestrian, with much experience of the countryside, and is thus more likely to be receptive to a question such as '*have you ever seen a big cat whilst out riding?*', or '*have you ever had any unusual sheep losses?*'

I once asked a member of the South Durham Hunt if they had ever put up a big cat, and she promised to get back to me on it - I'm still waiting for her reply by the way! Another customer was purchasing an extremely large bag of bird food, and I asked what was the most interesting thing they had seen at the bird table, and was told - a large black panther! In all cases I either tell them about, or scribble down, the Big Cats in Britain website address, and ask them to take a look. Sometimes I really stick my neck out, and leave my 'phone number with them, and ask them to get back to me should they see anything. Surprisingly this does produce results, and I have had one or two good sightings from this method.

With the home shopping vans you have much greater freedom, and the people - particularly those out in the farms - seem to be a lot more approachable, and several have a story to tell, or know someone who has. My latest report from Newby near Middlesbrough was obtained in this way, and I'm sure I'll acquire several more as I go on.

I have yet to see a big cat, but getting out into the countryside on the home shopping vans increases my chances, so perhaps one day I will join that select few who have been lucky enough to see a large feline in the English countryside.

Big Cat Diary

Many people wish anonymity when reporting their experiences, and although we have their contact details we shall never make them public. Names that are mentioned in the following round up are those of witnesses who have no objections to their details being used. It must be made very clear that we are not endorsing any sightings or any photographs. We are presenting the round up here without comment.

England

Bedfordshire
2 reports

11th July: While security officer David Gater was sitting in his truck at 04.30hrs on the Thureigh Air Field Business Park, he was surprised to see a large cat, which he believes, was a jaguar. The animal was black with a very long tail, and was moving at speed across the field. It had rounded ears and was observed for around ten minutes - the closest distance it came was approximately 20yds away from the truck. **(Source: *BCIB*).**

23rd December: *"I was driving back to Luton with three members of my family, and came off the A6 when I saw a woman cross the road by the bend near to the quarry. It was very odd as she was all in black, fur type hooded coat, black boots, and holding a carrier bag. I saw her face in the car lights, she put her head down and was very pale, she had blonde hair and was tall. Seconds later a huge cat came out of the fields behind her, causing me to slow down in amazement. The cat bounded up to her and they both went quickly into the quarry. We stopped the car not believing what we had just witnessed. We looked through the bars of the quarry gate but saw nothing. The animal was definitely feline, it had a collar with something shiny on it, the woman called to it and it arched its back like a cat does when pleased and the woman was not afraid of it.*

The cat was huge like a big dog-size but it was not a dog, the tail was long and the legs were short and it had a cat-like gait. It was very black and had big paws, green /yellow eyes, long tail almost whip-like and it was flicking. It had short legs and was stocky, with pointed ears. We watched for about 3-4 minutes. Its height was 3ft, and its length around 4ft. When it bounded, it did so with the effortless grace of a feline.

Friends say there is a local legend of a white Wicca witch who came across a panther in a quarry some years ago and tamed it. The witch, panther and its cub roam the quarry and the surrounding fields. All three are seen together at the winter solstice the cats being either side of the woman.

The same woman has been seen many times, with a panther, in the DSM Quarry by Millers Way. No one has seen her face and she speaks to no one - she appears and then disappears. There is talk she uses underground road drains to make her escape. She knows the countryside well and knows of the catacombs. The same woman has also been seen with the panther along in the Green Lanes by Sewell Village and by thorn cutters on the A5. Local rumour has it that the panther is a female pet, and wears a jewelled collar.

The cat seemed delighted with the woman, and whatever was in her carrier bag. Its tail flicked about like a whip and it arched its back rubbing its head on the woman's arm. The woman was more interested in getting away from us. It was like something out of a movie, a woman all in black with a hooded coat like a monk, and this unbelievable blue - black cat. The cat was obviously used to people as we were in the car only a metre away but it was only interested in the woman. The whole thing was very eerie, sure as hell no one would attack her in that quarry." **(Source: *BCIB*).**

Berkshire
9 reports

April 2007: Ian Duffin, a dog handler in the army, came face-to-face with what he described as a black panther "the size of a German Shepherd" in Shinfield. He spotted the animal leaping a across the road only yards in front of him, while on his way to work at the Arborfield MoD base. **(Source: *Reading Evening Post*).**

May: Patricia Lewis spotted an unusual cat and told the *Reading Evening Post*: *"It was the other side of the river from Pangbourne, out towards Ashampstead. I was driving up the valley road when something caught my attention. I thought, 'my God, what's that?' and stopped the car to get a better look. It had a big body, a small head and a very long tail."* **(Source: *Reading Evening Post*).**

May 2007: A sighting occurred on the A3095 between Jealots Hill and Bracknell, heading south towards Bracknell, just before the junction with B3034. The creature was black with no apparent markings. It had rounded ears, but the witness did not notice the tail. It was described as a large unusual feral, and was spotted for about a second, at a distance of around 200yds. The witness reported: *"I was driving along the road at 7am when something caught my eye. It was a black cat loping along a hedgerow. I saw it for perhaps a second before a roadside hedge obscured the view. My immediate thoughts were that it was not a domestic cat. My view was so brief that I really cannot tell you anymore about it.*

I can only assume they are large feral cats or escaped big cats. They don't seem to do any more harm than foxes." **(Source: *BCIB*).**

24th June: Sighting on the Bath road (A4) at Burchetts Green, (between Maidenhead and Reading) at 04.00hrs. The witness said: *"From what I could see, the cat was greyish with some darker markings, less markings than say a tabby cat. It had a long tail, which seemed to be well balanced, and it carried it almost* *horizontally from the rear as opposed to drooping down.*

I was driving towards it, and spotted it at about 30 - 40yds away, but drove right past it to my left as it crossed the road. It was in view for about six seconds. I was driving along the A4 from Maidenhead in the direction of Reading. The Burchetts Green area is fairly rural with woodland, farms and in fact a great deal of land used by the agricultural college. There is a roundabout with a sneaky road coming off it to the right and I am always wary to slow right down, as it's easy to be caught out there. Before I approached the roundabout (60yds way), I noticed the head of a cat with something in its jaws on the roadside to the right coming out of the woods, it trotted across the road in front of me and headed off into the fields on the left." **(Source: *BCIB*).**

14th July: *"I was playing golf at Lambourne Golf Club, in Burnham Beeches, Slough. I was with three friends, and at about 09.45hrs, from out of the bushes, came an ocelot. At first we had no idea what it was. It came to within 8ft of me. It was not afraid, but seemed distracted. I think it was looking for a way over a nearby fence that leads back in to the Beeches (very large wooded area 5 – 10 miles square). It stayed around us for about 40 seconds, and I got very close to it. The Internet confirmed it was an ocelot. As far as I'm aware, no other members have seen the animal. The four of us that day saw it very clearly from 8 ft away."* **(Source: *BCIB*).**

7th August: Sighting at old disused cement quarry at Houghton Regis at 15.30hrs. It was black, with pointed ears, and a long tail. The animal was seen for five minutes duration, at a distance of 20ft, and was similar in size to a medium-sized dog (Labrador or small Alsatian).

"I was out in deserted quarry walking my puppy, when I saw in the near distance what I thought was an unaccompanied black dog lying sunning itself. We were walking downwind of the animal at the time, and as we got closer it looked up, saw us, and when it stood up it was very clear that it was a cat. Its tail curved downwards and swished like a domestic cat. It walked very calmly to a nearby group of brambles and shrubs, and it sat there watching us, and as we retreated it turned and disappeared into the undergrowth. My puppy wet itself, and his hair was upright and he was whimpering - he is used to playing with cats and other dogs.

I had heard previously that there had been sightings of this animal, and used to think people had been seeing a large domestic cat. Not any more! I think this animal should be left alone, and my concern in reporting it is that it will start a frenzied hunt and the poor creature will be killed." **(Source: *BCIB*).**

8th September: Sighting at a village on the A330 (Ascot Road) called Touchen End at 2am. *"My house is a 2 acre farm surrounded by adjacent field. I saw a cat: grey, with lighter grey markings. Its tail was longish and pointed up-*

wards – it was over 1 in in length. It was a large unusual feral and was very quick. I saw it from less than 10ft away. I was shutting a window and it ran across the front garden at unbelievable speed. The size and the speed was the most alarming. It was at least a metre in body length, but it was hard to tell the height - similar to a medium/large dog.

I was in the front room watching a movie when my intruder light went on. I got up and looked out the window, and this is when I saw this thing. I have a lot of wild rabbits on the property, and animals, like foxes for example, are spotted quite a lot." **(Source: BCIB).**

23rd November: Geoff and Sylvia Killgallon, from Caversham Park Village spotted a large black cat with pointed ears and orange eyes. Mr Killgallon reported: "It was in a field out towards Dunsden Green. It was so big I do not think it could have been a domestic cat. My wife was convinced it was a panther. It walked up the field swishing its tail before it looked round towards us. I do not think it was hunting anything but it looked through the hedge and a pheasant walked behind it." **(Source: Evening Standard).**

Mid-November: Witnesses spotted a large black cat near Pangbourne. They said it had a big body and a small head. **(Source: Evening Standard).**

Buckinghamshire
14 reports

January: Prints found in an undisclosed location. **(Source: Jan Williams).**

February: A dog walker said he spotted a large black cat on The Rye, High Wycombe, earlier this year. **(Source: Bucks Free Press).**

February: David Zaborowski in the Rye river area, High Wycombe saw a cat that was beige/grey brown, 1ft long, and quite fat. It had rounded ears, and was seen for about 10 seconds at a distance of 20m, and was 2ft at the shoulders and 4ft long. He said: "I was walking in the woods near the river Rye in High Wycombe, and found a quiet spot to sit and draw some pictures of the woods for a project I was doing. I saw something about 20m away from me to my right, and then it disappeared behind a bush. It emerged again two seconds later on the other side so I stood up. The cat, started, flinched and looked straight at me, then turned round and ran away." **(Source: BCIB).**

13th April: Sighting at Cadmore End at 13:45hrs. It was light fawn, with some mottling, and was paler underneath and had pale, pointed ears. It had a long, striped tail about half the length of its body, with the lower half held parallel to the ground. It was watched for about 1 minute at a distance of 50m and was 75cm - 1m in length, with a tail about half this much again. The witness said: "I was having a beer (just one, honest!) in the garden of the Old Ship (Cadmore End) on Friday lunchtime when I spotted a largish animal moving in the neighbouring field. There was a hedge next to me, but no leaves, so I could see clearly, while it could not see us (there was three in my party plus another group in the beer garden).*

My initial thought that it was a fox. This didn't last long as it was clearly a large cat stalking something, and actually considerably bigger than a fox would have been (although not, I think, big enough to have been a cheetah, and the colouring was wrong - see below). It was about 50m away and I did not have my bins, but there was no doubt that it was a cat and it was a LOT bigger than a domestic tabby. It had a very long tail (ruling out a lynx), which was striped and with the lower half held parallel to the ground, rather like a cheetah. It was overall a light mottled colour, paler underneath, and prominent pale (when seen from behind) ears. A trawl of Google images gave me this as the closest I could find (jungle cat): I'm sure it must be an escapee, possibly interbred with a domestic?" **(Source: BCIB).**

13th April: Sighting at 19.10hrs in Yeovil. It was clear, and cooling in the evening after a warm day. The cat was black, with rounded ears and was seen for 3-4 seconds, at a distance of 75-100yds. It was 4-5ft in length, and the witness said: *"At the bottom of our garden we have conifers which were cut back last Autumn, resulting in large stubs of cut back trees. These stubs are approximately 20-25ft high, and on these stubs we saw a large black cat-like creature, slim in build, with long, tiny ears. It stretched its body and looked around the area. We were sat on the patio eating our evening meal at the time and happened to look up and saw this creature. I went to get the camera, while Mr. P saw the animal arch its back, look around and disappear."* **(Source: BCIB).**

31st May: Sighting by Mrs Joy Knowles-Cripps from Aylesbury. It was a calm, warm sunny early evening, after some previously heavy rain. The colour of the cat was black, and it had rounded ears, with a tail of approximately 3ft. It was seen for thirty seconds to one minute, at a distance of 70 - 100m. She said: *"My first thought was, that's a big black lamb and then I saw its tail, I am finding it really hard to give a height and length, as I will explain below.*

I was walking my dog tonight at about 8.15pm in a farm field near my house, which runs alongside a brook (Bearbrook) and near to the Oxford road (A418) to Stone/Thame/Oxford. There are long grass/trees and hedgerows all around the field, with adjoining fields beyond that. There is a large badger set in the next field up.

I was walking along the top of the field with my dog on the lead, and as I looked ahead I stopped as I saw a shape of an animal sitting facing me, very still. I was puzzled as I couldn't see anyone else in the field, and at first thought it was a dog or very big rabbit (I'm not mad, I just couldn't work out what it was sitting there, so everything goes through your mind). So, I stopped walking and just watched it. As I stood there it jumped up and over into the long grass at the back of the field, my first thought then as silly as it might sound was "that's a big lamb ?!" (There are no sheep in the surrounding fields) but as it jumped I saw its long tail, oh yes and a crow was agitated by the animal and swooped down at it as it jumped. I was then quite worried, and turned round and walked back in the opposite direction. When I got home, I told my husband, I am convinced it was not a domestic cat and it could not have been a lamb! - I thought I must sound crazy.*

I have been worrying about it all day, and am concerned as it is a popular field for dog walking, and near a residential area. I am naturally concerned now about walking my dog in the field. Should I be? Should I notify the local police? To be honest it has frightened me, and I would feel awful if I then read in the paper about an attack by a cat and I had not reported it, as I understand the police are not always very understanding with issues like this and think you have nothing better to do!

I would really appreciate your further advice on this, or should I try to forget about it? The more I think about it, I am convinced the cat was laying down watching me walking towards it. I must have walked passed it first with my dog before I turned around to walk back, which is when I then saw it. It has really spooked me!" **(Source: *BCIB*).**

Early June: Trevor Smith investigated a report in which a domestic cat was attacked. When he arrived at the house in High Wycombe, he found a scratch pad with marks that were 4½ft off the ground, which he believes had to have been made by big cat. **(Source: *Bucks Free Press*).**

Early June: A large black cat with a "big head" spotted near the Hazlemere golf course. **(Source: *BCIB*).**

21st June: Two women were out walking near Bisham at around 2pm when they spotted a mystery cat, which they describe as being the "size of a small pony." **(Source: *Bucks Free Press*).**

1st July: A couple near Hazlemere spotted a large mysterious cat and rang the *Bucks Free Press* who reported: *"A large cat-like animal was seen by Suzanne Young, 19, of Quarrendon Road, Amersham, and her partner Matthew Tucker, 19, of Orchard Lane, Amersham, as they were driving back along Amersham Road from a night out in High Wycombe.*

Suzanne said: *"All of a sudden a very big cat trotted out in front of our car from one side of the road. We had slowed down for a bend and saw a black blob on one side of the road. At first, we thought it was a fox until our headlamps hit it. Once this happened, it sprinted the rest of the way to the other side of the road.*

When we saw it in the light we realised it was far too big to be a fox, or even a dog. We weren't drunk as we were driving." **(Source: *Bucks Free Press*).**

19th July: John Almond from Penn Wood saw a big cat on the A404 between Amersham and Hazelmere. He reported: *"I'm a warden for the Woodland Trust & spend up to two hours a day walking my dog in Penn Wood. Earlier this week there was a report of a dead muntjac on one of the paths, which had been apparently 'attacked from the rear'. I took a few photos and moved it off the path so as not to cause distress for other walkers. Looking at it closely, it was badly bruised and I was happy that it was a road traffic accident. That evening I was walking with a friend and he wanted to see it. There had been a report in the Bucks Free Press a couple of weeks ago about a 'big cat' sighting on the A404. I put it in a memorable spot within 300m of my house, but when I tried to find the deer it had gone. We spent a good 20 minutes searching but could not find it, nor did my dog. There was evidence of something being dragged deeper into the wood. Last night I saw an unusual footprint near to the area, and have been back tonight to take some photos. Some are better than others.*

My mate said to get in touch with you, I'm afraid to say that I'm a non-believer, but comparing the print to a mountain lion print from an American site it looks pretty similar. I put my Ford key next to it - that measures 85mm as a scale. It was in the region of 110 x 100mm.

Three or four times in the last two weeks my black lab has refused to go into the woods; sitting down on the path, and no amount of dragging will get her in there....spooky ...." **(Source: *BCIB*).**

2nd August: Sighting at 07.30hrs on the Missenden Road, Butlers Cross. The colour of the cat was yellow, with a long, hung down tail. It was only spotted for a matter of seconds and was bigger than a puma – witness drove past.

"It came from the side of the road from a wooded area and walked very slowly down a slope towards the road - not interested in traffic. I was driving at the time. I'm very sceptical about what I saw. It seemed very large and I am sure someone else would have seen it if it was there." **(Source: *BCIB).***

4th August: Seen at Tamar Close, Loudwater at 02:30hrs. Animal was black with rounded ears and a long, sleek tail and was the size of a full-grown black Labrador dog. The sighting lasted five minutes at a distance of 6 ft. The witness reported: *"Mom was visiting and sleeping in the lounge. Front windows of the house were open because it was really hot that night. Mom was woken up and went to look out the window and saw an animal - as described - resting on the front lawn. The animal was constantly looking around, standing up and sitting down, as if assessing its surroundings. Didn't seem domestic, and the animal was on it's own. She watched it for about 5 minutes then left the window, came back a few minutes later and watched again to see if it was still there, which it* was, and then she watched for another few minutes. She noticed that the fur was really shiny and black, which was unusual. I hadn't heard of the other sightings and happened to mention it to my partner about her seeing something and he commented that he had heard of puma sightings in the area before. So we looked on the internet to see if we could find something out." **(Source: BCIB).**

28th August: At Weston Underwood, a large black cat was seen. It was much taller, longer and more graceful than a fox or a dog, with a long black tail, and green eye reflection in the lamp glare. The tail was as long as the cat (perhaps longer) and very thick, and static in hold. It was spotted for approximately two minutes, at a distance of 200m. It stood above the second rung of barbwire, perhaps taller than a Labrador.

"We were rabbit shooting, in a Discovery vehicle, on a stubble field, working down a hedge line. I was shooting ahead and my colleague was shooting from the passenger seat across to the left, using silenced air rifles.

I saw movement on the fence line and immediately knew what I was looking at!! I swore!!! and told my friend to look, he also swore (more than me, he's a Londoner!!!!) we were transfixed and couldn't move. We observed it for a minute more before it stepped through a depression in the fence and disappeared down into a rough hilly field! I was privileged to witness this. I see barn owls, badgers, buzzards etc everyday, but this was special!" **(Source: BCIB).**

Cambridgeshire
12 reports

29th January: The witness, at 10.45pm, spotted a mysterious cat on the old A15 between the Werrington and Glinton roundabouts, Peterborough. The cat was dark brown and was spotted as he drove by - it was sitting, waiting, while he passed to cross the dual carriageway. The witness noticed it had rounded ears, and was about 1.5m tall. **(Source: BCIB).**

4th July: Raymond Franklin, from Wimpole, who saw the cat at 10:15pm, said *"I'm sorry but the sighting was over in a second and with not the best of visibility. I could only see the bulk of the animal and the movement from 30yds (through rain).*

The cat was the size of a largish dog ... but seemed slightly longer at stretch.

Last Wednesday evening I was approaching the Bassingbourne roundabout on the A606 (going east to west) when, roughly 30yds to my front, I saw a large animal run across the road and through an opening into a nearby field.. It was a wet evening, and my view was very brief. The animal was dark and roughly the size of a Labrador dog. My first thought was at it must have indeed been a dog from a nearby house, but the speed, and the movement of the animal, bore no resemblance to the way a dog would move, and its speed was way superior. I tried to look for it in the field as I passed but there was no sight." **(Source: BCIB).**

14th July: Sighting at Ware Lane, Houghton, Huntingdon at 21.00hrs. It was black, with the height, length and length of tail - 2ft level to its back, from nose to bum 3ft, tail roughly 2ft.

The witness said: *"It had a roundish face, and was pure black with no other colourings"*. And they also saw a domestic cat roughly half the distance from the black cat which gave them a rough size and the black cat had them same characteristics as a domestic cat. The witness continued: *"This is a sighting by my parents on an evening walk on their holiday."* **(Source: BCIB).**

2nd August: John Page was visiting friends in the village of Cottenham when he saw a strange animal in broad daylight, (13.00hrs). He said: *"I saw this animal walk from a field entrance across the road. It had a long tail that hung down and turned up at the end. It was dark coloured, it had cat-like features and moved like a domestic cat."* **(Source: *Cambridgeshire Evening News*).**

8th August: Sighting at Castor Hanglands, 4 miles west of Peterborough, 10.00am. It was light sand in colour, with **a** long tail (almost the same length as the body) which was of uniform thickness, and with a rounded end with whitish tip. The animal was about the same size as a Labrador dog and was seen for less than 1 minute, at about 50m away.

The witness said: *"I was walking my dogs, on leads, on a path with thick vegetation on one side and open woodland on other. I saw the animal and stopped to watch as it was certainly not a fox or deer, which I have frequently seen. I could not see the head clearly, as the animal was nosing around a bush. It did not appear to see, or scent us, and moved into woodland. There is plenty of deer in this area. I returned today to look for "evidence" but found none."* **(Source*: BCIB*).**

10th August: Sighting at 7.55 am by Anne Raiment and one other witness (her husband without his glasses). It was dark brown, with thick velvet like fur and she could see light through ears sticking up like triangles. Its tail was about a foot long, curved and sleek, not bushy. It was only seen from the back end, for 10-20 seconds at around 35 - 40ft, and was the size of Labrador. Mrs. Raiment said: *"We were emerging from house and looked out over drive, and saw the creature (we obviously disturbed it by opening large front door) and it sprang, with two leaps, into the woods opposite. I stood and listened, but could not hear any movement. I walked toward the edge of the road, but saw and heard nothing. I then had to leave to get to work."* **(Source: BCIB).**

11th August: Sighting at 21.15hrs in Cantelupe Road, Haslingfield by Terry

Brown. He said it was difficult to be certain of the colour as it was dusk, but he believed it to be dark brown, and definitely not black. He said it was about the size of a Labrador with a long tail, and it was seen for approximately 2 minutes at a distance 60-70yds. He said: *"My wife and I were looking over a fresh cut field at the bottom of our garden,when the animal came out of hedge and tree cover 20yds into the field. Before the animal disappeared, one of our Labradors came down to the bottom of the garden barking ferociously -, this dog does not normally bark at anything!!"* **(Source: *BCIB*).**

16th August: A large mystery feline was spotted along the A14 in Elsworth. **(Source: *Cambridgeshire Evening News*).**

23rd August approx: This image was taken from a home in Snailwell, which backs on to Chippenham Park. The picture taken by Ben Coles, who believes the cat is about 4-5ft long and 3ft tall. He said: *"It walked very low in the grass and took almost 10 minutes to get across the land at the back of my house.*

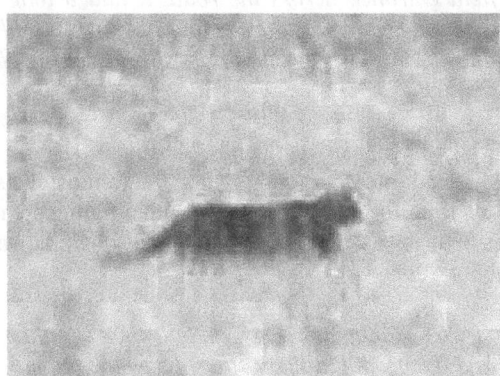

Ben was with his brother and sister-in-law in his kitchen when the big, black shadow caught his eye. He estimates it at 4-5ft long and 3ft tall.

He said: *"I tried to get some digital camera footage, but my camera wasn't working. And we did think about going out and chasing around the village in the car to where it would come out, but we thought we had enough evidence."* **(Source: *Cambridgeshire Evening News*).**

16th September: Seen by Terry Foster in Coln at 13.00hrs. The cat was ginger in colour, and he could not make out any markings, as he saw it from quite a distance. The tail was just under 2ft and its height was about 3ft by about 5ft, including tail length. It was seen for around five minutes at a distance of about 30m.

"We were driving down a really bumpy dirt track on our way to a lake to go fishing. I was the passenger, when I saw, in the corner of my eye, this large ani-

mal down another dirt track, which comes off the track we were driving on. I made my mate slam his brakes on and reverse. We both got out of the car and agreed with each other that it was something out of the norm. We both know it was too big to be a dog or a fox. To me it looked like a lion cub or something. When the large cat noticed our presence it casually stood up from a sitting position and walked away -, it was certainly not in a hurry. My friend and I watched it until it got out of sight, and we carried on our way. We would have attempted to follow it, but our car was blocking the track, and the track that the cat-like thing was on was locked with a gate." **(Source: *BCIB*).**

29th September: Sighting in a park on a disused farm, outside of Gamlingay, Station Road. The cat was jet-black, even with sun shining on its fur, and the eyes were a piercing light green with very narrow pupils. It had pointed ears and a long tail, just short of the length of the body. The back of the cat stood to approximately 45cm off the ground when walking. The first sighting was 5 minutes; the second sighting was 20 minutes.

"I was looking through a room in the house when I looked over the park and saw an unusually large cat rolling in the grass approximately 150yds away. The cat then walked to the right side of the park with something in its mouth, a rabbit, mouse or pigeon. It then rolled around the area it was sitting in. I grabbed my camera and attached a 500mm mirror lens to take some photos and opened the window of the room I was in and started taking photos of it. After it had finished eating, it moved to the lower field. I decided to go out with my camera to try to take some closer photos, which I managed to do. After moving to a third location to take photos, the cat ran towards a brook away from me and I lost sight of the animal.

I didn't hear any noise whilst it was killing an animal as the window was shut. I didn't hear anything when the window was open whilst taking photos." **(Source: *BCIB*).**

7th October: A two witness sighting at Kings Ripton Road, Huntingdon. The cat was black with a very long and curved tail, and was about the size of a St. Bernard. The sighting duration was eight minutes, at a distance of 300yds. One of the witnesses said: *"We were driving past a field when I noticed a dark shape in a unusual position. It seemed to be pacing forward slowly as though stalking, then sprinted forward with head down. The actions were very fluid, and it seemed as though the only part of the cat that was moving was its shoulder blades, which were moving up and down. I made my partner turn the car around at this point, as trees and bushes were now obscuring our view, and when we returned the cat was just sat there in the middle of the field with something in its mouth. It seemed to be staring in our direction, but by now it had got quite dark so we could not see what was in its mouth. Rabbits and pheasants are in the area, and the cat seemed to be making a clicking noise."* **(Source: BCIB).**

Cheshire
9 reports

26th June: Seen in Runcorn at Windmill Hill, verging on Bluebell Woods, near the bus-way to Norton Priory. The cat was black in appearance and the witness thought it to be a large unusual feral. The tail was very long and thick - thicker than its legs. It was seen for about 5/6 seconds, at a distance of 3-5ft. The height was approximately 1m to 1½ m, with its length proportional.

The witness said: *"The cat-like creature was very big. Its legs seemed to be very long, but thin, with large paws. Its claws were large and visible, and could be heard dragging on the floor. The animal was very thin and its face looked like a thin bullmastiff. I would say its legs were twice the length of an average cat, and its tail end was bulbous like. The animal appeared to be just as afraid, and turned, and then ran off into the residential area of Westwood on Windmill Hill."* **(Source: *BCIB*).**

12th July: Inspector Phil Hodgson, of Frodsham Police received a report of a mysterious cat in the area, and said: *"A man reported seeing a panther in the reservoir area near the Mersey View and Crowmere Lake. He said it was 10ft long and was running at about 40mph."* **(Source: *Runcorn Weekly News*).**

1st August: A sighting at 10.15hrs by Paul Twigg, in Wirral, on Puddington Lane, between Burton to Puddington. It was sand/beige in colour, with rounded ears, and a tail which was ⅔ of the body's length, long/thick to a rounded end with a black tip. Its height was 0.5m, and length was 1.5m (head to tail) and was seen at a distance of 50yds.

"I was driving from Burton towards Puddington in my works van. Ahead on the road, a very large cat (medium sized dog!) crossed the lane as I came to a corner. It was gliding cat-like, and jumped over a wall with ease into a garden. Its movement was smooth/light footed, and its long tail hung cat-like. No, but when I saw it I thought 'that's not a domestic cat'. I stopped at the corner were it had crossed and looked into the garden but it had gone. From seeing and stopping must have been no more than 10 seconds."

<u>**Alan Fleming reports:**</u>

This location fits in perfectly with my pet theory ever since the Bidston golf course video. There is a perfect route through the peninsular following railways, open land, water courses etc. Put all the Wirral sightings on one map and it forms a lovely straight(ish) line across towards Chester. The area it is covering butts up to the area (Runcorn Frodsham Delamere) where the black has been seen. I will be following this one up. The common route that seems to emerge lends itself to some "clutch points" where a sighting may present itself. **(Source: *BCIB*).**

18th September: *"Just after midnight I was returning home from a noon shift in Cheadle, travelling towards Leek on the approach to the main Leek/ Cellarhead road through Wetley Rocks. What I can only describe as a black panther type animal crossed from the right to the left of the road in front of my car. I had to brake sharply to avoid hitting it, as it came from nowhere and vanished as quickly, after slinking across the road and into a driveway. It was larger than an average Labrador in size and had a huge black tail. This will be the second time I have seen the cat in full, the first being on the edge of Alton Towers. After speaking to a colleague at work, he suggested I go on the website and report it to you."* **(Source: *D S*).**

31st September: Sighting one mile before Jct18 southbound on the M6 at

Holmes Chapel. The cat was black, with a very long and thick tail, and was about the size of a large dog - 7ft - and was seen for about 10/15 seconds at about 100m away. The witness reported: *"I was driving my lorry at 07.00hrs, when I looked into a field and it was stooped head down next to a large tree. It looked like it was drinking."* **(Source: *BCIB*)**.

18th December: Jet black cat seen near, or in, Tarporley (witness unsure of the Arriva bus route- bus number 84) at 3.45-4.00pm. Its tail was very long, and seemed slightly larger than at the bottom. It had pointed ears and was huge. The witness can only describe it as being the about the size of a Great Dane, and said: *"I was on the bus and it was not moving very fast. As I turned my head to look out of the window, the sight of a large black object on a lightly coloured house caught my eye. I immediately realised it was what appeared to be a very large black cat. As the bus carried on moving I turned and leaned towards the window for a good look, and I saw what appeared to be a huge cat-shaped black creature on the edge of the roof -, the tail was very long and, though moving, appeared to be draped down the edge."* **(Source: *BCIB*)**.

Month unknown: A grandmother from Halton said she saw a big black cat on school playing fields. **(Source: *Runcorn Weekly News*)**

Month unknown: A dog walker spotted a huge black cat on Halton playing fields (twice). **(Source: *Runcorn Weekly News*)**

Cleveland
3 reports

10th March: Sighting at Tees Barrage at the water treatment works, near the A19 flyover, at approximately 5.30pm. It was reddish gold, but too far away too see any markings. The tail was slightly longer than the body, and was held straight out behind as it ran. It was a large unusual feral, seen for approximately 3 minutes at a distance 200m. It was larger than a fox, fairly stocky, and its body was longer than its height. The witness reported: *"We were walking the dog (a small German shepherd) and we came to the small beck. There are normally hares and rabbits for her to chase along the beck, and we saw what we thought was a group of rabbits in a hollow. The dog notices and starts to run towards it. As she gets closer we realise it is a larger mammal of some sort, possibly a fox. We try to call the dog back, but she cannot - or will not - hear us. She is nearly on top of it before it moves away at high speed. It did not move like a fox, and its face was shorter and less pointed than a fox. It ran into an industrial yard, and the dog came back extremely excited. Compared to my dog, which stands at 23in at the withers, I estimate the height of the animal to be around 15in.* **(Source: *BCIB*)**.

22nd October: Sighting reported in the Stillington area. *"One of my friends has just had a sighting this morning, that my good friend Chris Hall will be very interested to hear. I have all confidence that this person would not lie, and I know he does not exaggerate. He says he was fishing at the park behind the church in Stillington, when, on top of the hill, he saw a large (too big to be domestic) black cat, sat on the incline, which slinked off into the distance. He gave this description of the animal: orange / yellow-ish eyes, slightly pointed tail. He kept remarking that earlier he had seen a man with a tan greyhound walk over the hill, and the cat was larger than that. He left the park soon after. I have enquired with Cleveland Police, who have unofficially said that yes there has been one report in Stillington." **(Source: BCIB).**

23rd November: Sighting at Sunny Brow, High Hesleden, Hartlepool Cleveland. The witness saw its profile and said it was the size of a large dog (about the size of a Doberman or Labrador). It was a large black cat with a large and long sweeping tail, and a square face. The witness explained that the large neighbouring back garden and its trees are near a cornfield and forest, and the cat was standing in amongst these trees.

The witness said: *"I was opening the bedroom curtains overlooking the garden when I saw the cat and called for my daughter. The cat stood there for around 5 minutes and then wandered back to the field. There was no noise, but our dogs have been barking constantly lately, and as the cat walked past other gardens there, dogs were going wild. I tried to get a picture on my mobile, but it was too fuzzy to see. My husband remembers seeing something similar in the same place a couple of months ago. In my opinion, this was a black leopard or panther. We are more wary about letting our dogs in the garden as it looked as though it was hunting."* **(Source: BCIB).**

Co. Durham
9 reports

1st February: Sighting by Gemma Prest, at approximately 5.40pm, at Lower Willington / Spennymoor in the Whitworth Hall Hotel area. The cat was black with pointed and tufted ears, and had a long tail and had quite hairy, silky fur. It was seen for about 25 seconds, at a distance of 20ft. She said: *"It was a bit bigger a Labrador but extremely long - 1½ times the length of my Lab. The cat crossed the road about 50yds ahead of the car. As we pulled along the side of the farm track in which the cat was going we got a good look, but the cat was just wandering away."* **(Source: BCIB).**

May: Sighting at Riding School near Hartlepool – witness saw large black cat apparently stalking horses. **(Source: Chris Hall).**

May: Black Cat seen crossing road at Sedgefield, in front of witness' car on the way to work. **(Source: Chris Hall).**

24th May: Brian Wardle was taking a stroll along the banks of the river in North Hylton when he spotted a huge cat 50 - 60 ft away. He said: *"It was the size of Labrador – a big black cat. It was just sitting on the path staring at me and I was frozen stiff. I didn't dare move."* The animal continued to stare at Brian for the next two to three minutes, until eventually scampering off into nearby undergrowth. **(*Sunderland Echo*).**

4th June: Sighting by Robert Chapplow, on the old railway line adjacent to Penshaw Monument, which used to be the old Leamside line. The nearest villages are Penshaw and Cox Green, adjacent to the A183 Sunderland to Chester-le-Street Road. It was a dry fair day of about 11°C and had been misty. The colour of the cat was black, with a long tail (length 2-3ft - like an S shape) and a small head. The duration of the sighting was 30-40 seconds, and the height of the animal was 3ft, with a length of 4ft. Mr. Chapplow said: *"I was cycling along westwards toward Penshaw on the old railway line now used as a cycle track. I was going round a slight right hand bend when I saw this largish black cat-like animal in front of me just strolling down the track. It then stopped about 50yds away from me, and when it saw me it slowly did a U-turn, strolled back about 20yds and did a left turn into dense trees/shrubbery. When I went past it, it made a low growling type of noise."* **(Source: BCIB).**

June: Ian Bond reports: *"I have just had a call from a chap from Hartlepool who was out jogging this morning at around 10am on the Hart to Haswell Walkway, which runs North West from Hartlepool. Just before Hesleden, he saw a cat's head poking up out of the grass on the verge. He estimated he got within 30m before the cat got up and ran ahead of him for 20-30m before disappearing off through the grass again. He got a brief side view, and then a view from the back as it was running away. His description was black, 24in high to the shoulder. He compared it to the Border terrier that he had with him and said it was twice the height and three times the length. (Not your standard moggy then!). He even went into the grass where the cat had emerged from with the dog so he could make a fairly accurate comparison."* **(Source: BCIB).**

9th July: Cat seen by Paul Ryder at 9:50pm, near the Pennine Way Garage, just out of Evenwood. The tail looked to be about ⅔ the length of the cat itself, and it hung low to the ground, curling upwards. The tail raised and curled more as the cat negotiated the terrain. It was seen for 2-3 minutes, at roughly a distance of 150 - 200m. The animal was no less than a metre long, probably 1.2 or thereabouts (not including its tail) and about half a meter tall. Mr. Ryder said: *"I was driving home from a friend's house and came to the end of a new by-pass that was just opened the previous week. As I got to the roundabout at the end, I noticed a creature, which for a split second I presumed to be a big dog, but immediately after realised it was feline. I went around the roundabout once more and pulled up in a garage just off it, got out of my car and crossed the road to observe it. The cat was walking across a field with its head down, pausing every now and then to look around.*

As it got about ⅔ the way across the fields it hopped across a ditch of some sort and back over again. I own a cat myself, and can say this action was definitely of feline nature. Eventually the cat disappeared behind scenery.

There was no noise, but the field it was in also had horses, which looked quite spooked. They did not jolt as such, but I could see they were nervous. The cat did not seem to pay any attention to the horses apart from perhaps a passing glance. I always found the theory of big cats in Britain quite plausible and think the release of pets in the 70s is the most likely cause." **(Source: BCIB).**

20th October: *"Just south of Durham at noon on Saturday, October 20th a farmer was inspecting a sheep carcass which had, where its throat should have been, a mass of blood. It might still have had the head attached (just), but we could not tell at the speed we were travelling. The rest of the sheep were all crammed in to one corner of the field as far as possible as they could get from it.*

No further details, although by the time I next log in I might be able to pinpoint exactly which field it was in - I need my rail maps to work out which one it was. I think I have narrowed it down to an area of land between two settlements, but obviously would like to be more precise in case more than one farm covers the area.

It may have been a dog, or some sick individual, but given the mess it was in I thought it was worth mentioning it. **(Source: Rachel Lacey).**

Month Unknown: Eddie Davies spotted what he believed to be a large unknown cat on Seaburn Dene, yards away from occupied houses. **(Source: *The Sunderland Echo*)**

Cornwall
12 reports

13th April: Claudia Chalmers reported a sighting. She was just in Cornwall, but wasn't exactly sure where - Pensilva? The animal was black with a tail of 1 metre-ish, but she was not sure about the length, but said that the tail was a good ¼ of it though! She cannot remember the height either, but is pretty sure it was the size of a small leopard of some sort.

"I just came out of a small shop like a newsagents and I saw a large black cat standing alert in a field, it was quite hidden, I saw it in a field with very high grass - not very good camouflage - black and yellow. It flinched, and then it sped back into the distance, it was looking on the floor." **(Source: BCIB).**

21st April: Sighting by Elizabeth Mitchell at Viaduct Farm, Lower Penponds,

Camborne Cornwall at 8.50 am. It was black/dark brown in colour with long tail, rounded at tip. She saw it for three minutes at a distance of 75yds and said it was about the size of a fully-grown Labrador dog. Ms. Mitchell said: *"As I was going out of the door I noticed an animal on the farm trailer in the field walking along edge of the trailer sniffing between seed trays. It was not a dog or fox; far too large to be domestic cat. I went back indoors and grabbed my binoculars; it was definitely a big dark coloured puma type cat sitting on the trailer. I could see the profile of the head, square jaw lioness-like head. It then started to continue along the trailer. I then walked towards it to get a closer look, but unfortunately the geese started cackling, and when I reached the field it had gone"* **(Source: BCIB).**

9th May: A picture printed in the *Western Morning News* was said to be of a large black cat. Zoologist, and owner of Tropiquaria in Somerset, Chris Moiser, believes it to be nothing more then a squirrel caught at an odd angle. **(Source: Chris Moiser).**

May: Gill Bunker spotted a large black cat from 80yds away, which seemed to be stalking something in fields near Frome. **(Source: BCIB).**

3rd June: A two witness sighting at 5.45am, Liskeard (7 miles away) between village of St. Neot and Hamlet of Draynes during patchy hill fog and drizzle. It had a ringed tail, and there appeared to be circles on the body, which was a greyish colour. It had short fur, and a long tail which curved down. It was seen for about 10 - 15 seconds, at a distance of 5m - taller than a fox. As it was, a rear end view could not judge length. One of the witnesses, Kevin Wright, said: *"I was taking a friend, Becca Carberry, to work, when she saw the same, or very similar, animal about ½ mile from this location on her way to work yesterday morning, though she was riding her motorbike yesterday. We were travelling along a very narrow back lane with overgrown hedges - all grassland woods and hedges around, and moorland 1 -2 miles to north of the locations seen. The animal was in the lane as we came round the bend and ran along the road in front of us until a gateway enabled its escape. My speed would have been about 10-15 mph."* **(Source: BCIB).**

11th July: Sighting reported by Peter Roberts. There were three witnesses and he said: *"We were parked in the car park off Shrubberies Hill in Porthleven, 17:00. The animal was all black with a long tubular tail. Height and length: It would depend on the height of the grass in the field. But when it sat down it was half obscured, and when it moved across the field the bottoms of the legs were obscured. The sighting lasted for several minute, with the initial sighting being of something black in a yellow/brown field. I would estimate the distance at about 150 to 200yds (from the car park, across a road, a field and halfway into the second field).*

We were returning to the car having spent an afternoon at Porthleven beach. I noticed this animal in the field, which I brought to my wife and son's attention. It was initially seated. After a while it stood up and padded (it didn't run, it didn't walk - it stalked as my son described it) its way from the field. I did say that maybe it was a fox because the alternative is too weird, but I could not square that with myself at all. This animal was all black with a long finite tail. And at the distance seen it must have been a large animal (at our distance I would estimate the animal to be just under 1 centimetre - do not know if someone can back calculate scale from the distance). I wish I had taken a photo, but loading kids and beach kit into the car, and with an hour's drive ahead, it was not foremost in my thoughts! It was a clear, sunny day - the best of the week - and a direct view with no hedges/trees etc. to interrupt the view. The car park we were in is (or at least appears to be) higher than the field we were looking into." (Source: *BCIB*).

5th August: Andy, his wife Sarah, and two small daughters, along with their other daughter and boyfriend - the latter being on leave from the Royal Marines - took a walk on a "off the track path" between the villages of Foxholes and St. Stephens, in the China Clay District. As they approached the local river, they turned off the main track and headed for St Stephens. They came upon a small copse and in the middle was a large dead tree, no foliage, with two massive V shaped branches up above, the perfect perch! At the bottom of the tree was fresh bark that had been scraped off; they also found prints around the base of the tree. Something didn't "seem quite right", and they could not work out what kind of animal would have done this. Also on the other side of the tree something had climbed up it to a height of about 30ft to the V shaped branches, the marks were plainly visible. The tree had not been scratched, but something had removed the soft dead bark. They returned the next evening with a camera to take photographs, and they noticed that fresh marks and more bark had been scraped off. They are sending us the pictures via email.

It was a "strange night" for Andy, the clay pit area is a "spooky place" with shrubbery growing over old buildings; as they walked passed an old waterwheel Andy distinctly heard a woman's voice call his name in his ear! Andy's father, who was a Royal Marine, claimed he had been growled at several years ago in the clay works. He was quite unnerved by it, apparently completely out of character for him; he refused to go back to the area on his own for some time. Rabbits plague the area, so food would not be a problem for a large cat. Andy said that everyone knew about the cats in the area, but they had been quiet for some time and had not been reported. Work has recently started in the clay pits and he wonders if this has brought the cat out again. (Source: *BCIB*).

24th August: A large cat spotted sunning itself near the Dragon Centre in Bodmin. Larger than a Labrador dog, around four ft long and pale in colour. (*Cornish Guardian*).

24th August: Sighting at Bodmin. It was sandy/fawny and looked like a puma.

It had a long tail, and pointed ears. Its height and length were 30in and 48in respectively, and was seen for approximately 15 minutes. The witness said: *"I was sat in my car in the car park when I noticed it walking the edge of woodland. I could not believe it so looked at it through binoculars, which confirmed it, it was a puma not a dog. I watched it walk along, then it stood for a while staring, then sat down and eventually lied down. The cat bounded away into the trees when it caught sight of a human who spooked it."* **(Source: BCIB)**.

27th August: Telephone call received from a man in the St Ives area of Cornwall. Found what he believes are lots of tracks all over the place. **(Source: BCIB)**.

5th September: Sighting in Bodmin by witnesses who had moved into their property in June, when their dogs howled as soon as they entered the garden, and started digging in one spot. The dogs play up every night/early hours. There is a 7ft fence surrounding the garden, but a large animal visits the garden every night and leaves faeces, scratch marks on the trees, and a flattened area of grass. 11.45hrs on the date above was when they finally saw the large, black panther type cat. They rang Newquay Zoo, who gave them the BCIB number. **(Source: BCIB)**.

September: Report from Bodmin playing fields – a locked nature reserve. Peculiar sounds: *"motorised purring - teeth clattering - high pitched squeaks early in the morning starts sometime on a night at 4pm stops at around 11 am - intermittent throughout the night, not sure what it is."* **(Source: BCIB)**.

Cumbria
5 reports

27th April: Paul Caffrey, who was at the wheel of an X35 express bus, insists he was not seeing things when he spotted a large black cat run across the A590 carriageway.
He reports: *"I was coming from Kendal driving back to Barrow on Saturday morning on the X35. I'd just come through the Greenodd bends section of road and got to the first section of dual carriageway before it becomes two lanes. Then, all of a sudden, this big black cat with a very long tail ran across the road onto the central reservation. I slowed down and the passengers must have noticed, but I couldn't stop because there were about seven cars behind. It was about 1½m long. As I passed the animal, it turned and looked back at me and the lights of the bus lit up the cat's amber eyes."*
(Source: *North West Evening Mail*).

14th July: Sighting by two witnesses between midnight and 01.00 hrs, at Mealsgate, Cumbria. The cat was seen in the car's headlights and was said to be mottled brown.

"I am a special constable in at Wigton Cumbria. In the early hours of this morning, my colleague and I saw what we believe was a big cat crossing the road at Mealsgate. It was about the size of a large fox but did not have a bushy tail and definitely was not a fox. It was not a badger and it ran too fast. It was not a dog, it appeared to be a brownish colour. **(Source: BCIB).**

25th August: *"My name is Stuart Buxton, and I don't need to remain anonymous in this.*

It was Saturday 25th August and my partner and I were out walking. We were in between Rosthwaite and Grange, in the Borrowdale valley south of Keswick. Close to the B5289. At the time of the sighting I estimate our grid reference was NY 252 158.

My girlfriend and I had just climbed Castle Crag near to Rosthwaite in the Borrowdale valley of the Lake District. Our descent took us north to Grange and then we followed the river Derwent back south to Rosthwaite. On our route we came off the main path to be closer to the river, this saw us climbing a few rocks and treading less beaten paths. Approaching the top of some rocks in High Hows Wood we saw a small deer, she hurried off on hearing our footsteps, down the rocks and deeper into the woods.

Having followed her movement I was amazed to see a large black figure following in its footsteps 20 seconds or so later. I am certain by the shape and movement that it was a black cat. It was black, all over. I did not notice the ears but its tail was long, and thick, certainly that of a cat. It was approximately 2½ft high, and 4 to 5 ft long. I suspect it was a puma based on what I have seen in zoos and on TV.

There are lots of caves around that area so it would be ideal for hiding. I do believe it is not the first sighting of big black cats in the area, and it so happens that this sighting was only a mile or so away from the Wainwright peak Cat Bells. The name, which is rumoured to derive from "Cat Bields" which means the shelter of the wild cat.

I hope that this is useful, and that other people can confirm my sighting. They are a wonderful thing to see, I would love if there were more wild cats roaming the country. If this information is forwarded to local news, feel free to pass my details to them to further discuss." **(Source: BCIB).**

6th November: Large black cat - bigger than domestic dog and cat - running with legs stretched out seen at Brampton Townfoot Park, at 4.30 pm. It was very light on its feet, and did not make a noise. It was longer than a greyhound at about 1.5 m with legs out running, and was about 70cm tall. For about 3-4 seconds, the cat ran in front of the witness, about 10yds away.

The witness said: *"I was walking diagonally across the field behind my house, and I looked up to see a large black cat run across the field. It ran about 20yds then disappeared behind a tree. I went looking for it, but did not find anything. The cat made no noise. It was almost dark and it was hard to see due to a dark fence in the background. Last week I caught sight of a large black cat, but did not take any notice. A few weeks prior to this my cat had a very noisy terrifying fight in our back garden."* **(Source: BCIB).**

20th November: A lady and her husband were driving along the A592, having left Glen Rideln (?) and were approaching Penrith. As they came around the corner, they saw three large black cats running up the hillside and over the top, out of sight. The witnesses have never seen anything like this before and were not even aware of the subject, but after the sighting, they are fascinated and have been scouring the internet. The height of the animals was bigger than a fully-grown Labrador's but longer, and all of the three cats were roughly the same size. They ran "ever so fast" and were about 150yds away from the witnesses, the time was approx 16.00hrs. The car in front, who were the witnesses' friends, stopped and got out of the vehicle; they had also seen the cats. **(Source: BCIB).**

Derbyshire
10 reports

1st May: Lalla Blackden came face-to-face with a black cat-like beast while she was doing a spot of home gardening. She said: *"I was crouching down gardening when I looked up and saw this creature. It was the size of a big Alsatian. It was black and it turned when it felt me looking at it and stared at me with these big golden eyes."* **(Source: Mark Williams).**

14th May: Witness reported: *"Black, tail was the first thing I noticed (1½ft long). I was driving past (a passenger) 30ft away from the animal, which was 2ft in height and 3ft in length with a long tail. I was a passenger in a car and was looking at the scenery (on the Ambergate bottom near the Excavator Pub). I then noticed a large sized cat walking along the path. It was quite skinny and it looked like the hair towards the back end of the cat was patchy. It had a really long tail and was walking with its head down. We turned around at the next island to see if we could get a better view, but it had disappeared into the bushes. The sighting area was quite rural with fields etc."* **(Source: BCIB).**

23rd May: Sheep kills found in an undisclosed location in Derbyshire, believed to have been victim of a big cat by witnesses. **(Source: BCIB).**

July: Photograph of a 'large' cat walking along a dry stone wall, taken by witness Paul Grey near Great Hucklow. **(Source: Derby Evening Telegraph).**

30th August: Sighting at 11.45pm **by** Kim Redgrave at N Riverside Lodge, Whatstandwell. She said: *"It was very light, stripey, and very fat – the size of large ram -with pointed ears pitting down, and it was sitting down. Its length was 1m and was a large unusual feral. It was just outside the front door, and I was checking on the children and looked out the upstairs window. The cat was lying down facing the wall with its back to me."* **(Source: *BCIB*).**

5th September: About 2 miles south of Sumercotes, in a wood to the left of B600 a pure white cat with no other markings, was seen. It had a thin short-haired tail, turned up at ground level resembling a leopard's, and had rounded ears. It was about 18-20in at the shoulder, with a rounded head, the top of which was slightly below shoulder height. Its length was about 30in - slightly smaller than a black Labrador - but a lot thinner. The witness said: *"I observed it for 2 to 3 minutes walking slowly towards me as I was partially hidden by the bushes although the cat was in full view walking towards me, at a distance of about 80yds. I began to get concerned, as I have not accounted such a large cat before so stood on a branch snapping it to let the cat know I was there before it got too close. It looked at me and ran off into the woods."* **(Source *BCIB*).**

15th September: *(undisclosed location) "Yesterday I found my horse to have 20 neat slashes across its back end on the one side. We have very little barbwire fencing and the fence is only has two strands of the barbwire however the marks on the horse where in rows of three. I have checked the field for anything that would cause this and have found nothing. It was suggested to me that a 'big cat' could have done this and I am emailing to ask if a big cat would attack a horse and would it come back I don't believe the marks to be of human force as a horse would not stand the pain and would naturally run away from the pain, so this leaves me with little understanding of what else it could have been."* **(Source: *BCIB*).**

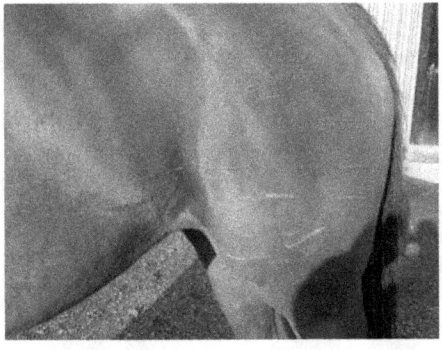

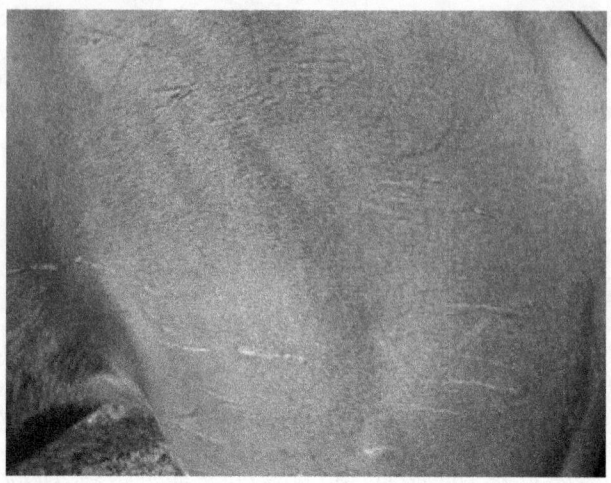

26th September: Guy Beston, and his wife Patsy, spotted a big black cat near Chevin whilst out for a walk down an old track. Mr Beston said: *"We were heading back towards the car, off the Chevin and it stopped in the middle of the track. It was smaller than a Labrador dog and totally black. It stood sideways to us, looked at us and then slinked off into the undergrowth."* **(Source: *Belper News*).**

September: A four witness sighting in Manchester at 12 midday. *"After walking under a rail tunnel and coming into the light I noticed, on a hill, an animal with a large stride come bounding over a hill, then turn and walk the other way to reappear moment later some distance away. I have included a photo of what I believe was a big black cat, although it is now some distance away."* The witness believes it was a jaguar. **(Source: *BCIB*).**

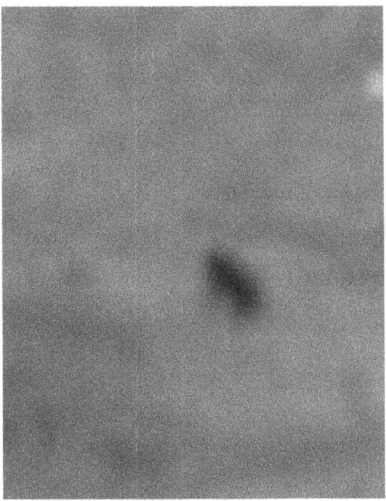

Date unknown: A couple visiting friends in Lea saw what they believed was a puma. Mr Hart from Leicestershire said: *"We were transfixed. It was the size of a large dog really but was jet black with a long slender body, small head and a long tail, which pointed slightly upwards. Its head hung low below its shoulders."* (**Source:** *Matlock Mercury*).

Date unknown: A woman walking at Caudwells Mill in Rowsley, reported seeing a tiger-like creature, strolling along the grassy slopes towards the river. She said: *"What I saw looked just like a tiger, with stripes of black but with browns and greys rather than orange colouring. The body was about 3–3.5ft long, with the tail adding another 2-3ft. The head was large and distinctive with the markings of a tiger."* (**Source:** *Matlock Mercury*).

Devon
38 reports

4th January: Alan White reports: *"I was travelling on the train coming back from Plymouth towards Newton Abbot. We were about 7 or 8 minutes from Totnes and at around 16.20 to 16.25 on the north side of the track I saw of what I knew was a big cat, black in colour. The tail was long and extended and above the rear of its body. The cat was facing slightly down the field with its head and face down to the ground. I said to my daughter did you see that, but she did not. Checking on the map when I got back home I am sure his was in the Rattery area. I telephoned Chris Moiser when I eventually got home and he told me that there is quite a lot of activity in that area. In my view it was definitely a cat, very large, panther like, very low body and very powerful. It did not move off the spot where it was standing. The only other remark I can make that the weather was fine and clear. It was an unobstructed view of the animal. As soon as I saw it the* natural instinct at first was to say 'no it can't be,' but immediately my brain took in the giz of the creature and came back as big cat." **(Source: *Alan White BCIB*).**

4th January: Witnesses reported seeing a puma at Stover Country Park. **(Source: *Alan White BCIB*).**

10th January: A puma spotted near to Ideford and Ashcombe. **(Source: *Alan White BCIB*).**

11th January: At around 8.15 to 8.20 there was a large black leopard seen looking somewhat bedraggled with the weather in the area of Whiddon Down, in North Dartmoor. The timescale in which the cat was seen was between ten to fifteen seconds. The witness was adamant she knew what she saw. I have spoken with the lady and she is going back to take a photo of the place where she actually saw the cat. **(Source: *Alan White BCIB*).**

17th January: Professor Jules Petty reports: *"I was in Devon on Tuesday - Wednesday. Within 5 minutes of arriving at my friend's house near Witheridge, a neighbour stopped to say she's seen two large black cats walking together near South Moulton."* **(Source: via *Rick Minter*).**

21st January: *"We were sitting in a lay-by at the side of the B3224 near Exford at 17.00hrs having a coffee and a sandwich when my wife suddenly said "oh my god! What's that?" I looked up and saw a large, and I mean large, brown cat sat up on a hedge bank. It sat there and then looked back down behind the hedge, then another big, but not as big cat appeared alongside, this one was black. I can only think that this was a mother and cub as the larger one of the two was licking the other as if to clean it. I went back yesterday and there were prints. We watched the animals for ten minutes; the tails were long and thick, the brown cat was about 2ft in height and the black cat was about 1 to 1½ ft tall."* **(Source: *BCIB*).**

29th January: Reports of possible puma sightings on the borders between Devon and Dorset. *(**Alan White BCIB**).*

6th February: 03.00am or so Bristol centre a black cat "far big to be a domestic, but smaller than a fox" **(Source: *Radio Bristol*).**

18th February: What was said to be a puma spotted near Dunkswell. **(Source: *Alan White BCIB*).**

26th February: *"Please find enclosed my letter which I wrote to the local newspaper and which they haven't printed. Since my first sighting of this animal, which incidentally lasted a good full minute and it was only 3ft away from me, thankfully the other side of the window, it has come back and was sitting* *actually on the windowsill outside growling and wailing. This was at 1.30am. on Monday 26th February 07. But when I pulled the curtain back a little to take a photo it jumped back onto the roof. On several occasions, my cat has alerted me to 'something' being out there.*

I have contacted so many people to try and get this investigated but no one wants to know, they probably think I am crazy but I know what I saw. I would please, like someone to contact me about this and if possible, to investigate as it is clear to me that this animal is still around. This is an urban area although there is plenty of wooded habitat in which it can live and hide."

Lynx in our midst!

"Has anyone else seen or heard the wild Lynx Cat in the Meadfoot or Wellswood area of Torquay? I live in Meadfoot Lane. Behind our properties, there is a 30ft wall on top of which there is an area of woodland. On the other side of this strip of woodland is Jacobs Ladder; steps which lead up from Parkhill Rd. to Vane Hill Rd.

I have a cat that goes out of my back bedroom window, jumps up onto my bathroom extension flat roof then across onto my neighbours back bedroom extension flat roof, across the adjoining wall to the woods for his little outings. He normally stays out between half to one hour. However, since the beginning of the New Year he has stayed out for 3 hours or more. During these long absences, I have heard loud growling wailing noises coming from the woods. As the woods are inaccessible to me, I have had to stand in my small back yard and call my cat from there. He still didn't come and I assume he had gone into hiding until the coast was clear for him to make a run for it. When he finally appeared he was clearly traumatised.

Last Wednesday, 7th February, my cat was sitting looking up at the window where he would normally go out. I approached the window and to my absolute amazement, there was this enormous cat staring through the window at me. This was no ordinary cat. It had short brown dense fur, very muscular body, about a third to half the size bigger than a domestic cat, its hind legs were long, about 8in from elbow to heel, it had a blunt (rounded ended) tail. Its head was large with pronounced cheek bones and large almond shaped yellow eyes which were unwavering in their scrutiny of me.

Its striking upright ears were huge with dark tufts of hair on their tips and were close together on the top of its head. Its mouth was a large upside down T-shape. It was growling and wailing with its mouth closed, it seemed to come from the throat but was unmistakably the same sound I had heard during the weeks before.

The next day I phoned Mr. Neil Bemment, the curator at Paignton Zoo who was not at all convinced and told me to take a photo of it! Unless it comes back to the window this is impossible since the wood has steep banks and is level with the roof of my house. I have since spent hours on the internet. Firstly looking at domestic cats. Could it possibly be? I thought. But nothing even remotely resembling what I had seen. I then went on to look at wild cats and Lynx's in particular - The coat, eyes, tail, height, length, and in particular the ears and cry, all add up to it being in fact a Lynx. The fact is that it lies in wait for its prey, pounces on its victim, breaks the vertebrae on the back of the neck and devours it. I believe it intended to have my cat for breakfast. Incidentally, my cat does not go out anymore, only into the enclosed back yard. Lynx's can travel up to a radius of 25 miles, but usually remain within a 3-mile zone if food is in abundance. This animal would be able to travel from Meadfoot Lane to Vane Hill Rd, into St. John's wood, across to Daddy Hole Plain and along the Coastal footpath to Hopes Nose or Ilsham Woods while virtually being able to maintain its cover in the vegetation for the whole distance!!! Living Coasts should be vigilant." **(Source: *BCIB*).**

9th March: *"Myself, brother and sister went shooting on the 9/3/07 it was very dark and we had a spot light, we got in a big field were there was a valley below us and wood my brother got out and got his gun from the front because we though they would be lots of rabbits because of all the gorse bushes around the field.*

"Then I sat on the front fender of the quad and my sister drove it up the field when we got to the cleave of the valley I got off and my sister screamed. I got the quad and raced forward saw this big black cat running from us. I thought it was a fox so I chased it. It stopped, turned round and stared for a moment turned then sprinted off. It was travelling more than 25 miles an hour because I was going that fast. It was leaping in front of us than I went round the hedge and looked up, it was there staring at me so I went back up the field to get out of its way, and it was running beside the hedge beside me. I quickly got out of there and went down to my house." **(Source: *BCIB*).**

April: eMail sent to Jonathan Downes & Chris Moiser - *"There's an article in this week's North Devon Journal (page 19) about a lady who witnessed a big cat and apparently has gouge marks in her shed door and a sample of hair."* **(Source: *Chris Moiser*).**

28th April: Report from Noah's Ark Zoo Farm near Bristol. *"Following a telephone call on Friday about Sheep kills at a farm park in Somerset earlier in the week I attended there on Saturday with Jane and Misty the dog (special permission was given for Misty to enter the park as dogs are normally excluded). The park had lost 6 sheep (4 found dead and 2 euthanized)) in 4 attacks in last July (2006). The killings started again in the last week (last week April 2007), 2 so far, the last one on Wednesday. One of the keepers has been sleeping in the field at night now to try to protect the sheep. He telephoned me having got the phone number from the website. I had not visited this place before. It is fairly new and the reason why I wanted to look at what might otherwise be an ordinary dog kill was because the sheep field was within a perimeter fence that was built to zoo licence specification, i.e. it is 5-6ft high 6in x 6in mesh and regularly inspected/ maintained.*

The attacks have been fast and quiet and the flock is left only slightly frightened. We took Misty (a Pyrenean mountain puppy) in when doing the field inspection and the sheep all trotted across to see her. There is much surrounding woodland - some owned by the local nature conservancy, and there are no general public houses nearby, just the odd farm. No pictures of the dead sheep were available but they were supposed to be very badly damaged, to the extent that the sight of them upset the keepers who found them and on 2 occasions shot the injured survivor. There seems to have only been one kill per attack, but on two occasions there was one dead sheep and one badly injured. The rest of the flock was fine. No evidence was found at the site at all, including any dog droppings on the surrounding footpaths. The fences were some of the best that I have seen in a long time, with possibly one badger and one rabbit scrape through into the field. The keeper is going to continue to sleep on the site at night, but at a new position, and call straight away if there are any more kills. He is also talking with the local farmers." **(Source: *Chris Moiser*).**

8th May: Thomas Skinner was hiking on the moors when he spotted movement. *"I looked and there appeared to be a large black animal roughly the size of an Alsatian, which I think, may have big cat."* **(Source: *BCIB*).**

May: Anthony Bevan and Christopher Johnston were contacted by a lady who lives on Exmoor and has had a number of lambs attacked and killed. A big cat has been seen in the field and also seen eating a lamb. They set two remote cameras up there to try and see what the animal was that has been attacking the lambs.

On one of the photos, it shows eye shine from an animal. What is interesting is that we got no photos of deer, fox or badger and we normally do if they are there. There were also no tracks of any of these animals. Whatever this animal is, it is very curious and aware of the camera, it seems to be moving slowly towards it close to the ground, close enough to set the flash off, but not to see the animal. **(Source: *Christopher & Anthony Bevan*).**

5th June A large black cat spotted at Lower Langaton Farm, Chittlehampton. **(Source: *BCIB*).**

9th June: Big Cats in Britain received the following email: *"I am a professional falconer and was flying a hawk on an area of Dartmoor with some American clients, when one of them pointed out this creature, it was walking along a path about 200yds away from us, it was black/grey, best size comparison is miniature pony. It had very thick shoulders, a long thick tail with a blunt end with small round ears. Its movement appeared feline, then bear like sprang to mind. There was a party climbing on the tor opposite making a racket and this it ignored completely. The clients took some photos, which are attached, they are not brilliant but should be of some use.*

I have worked with dogs all my life and it was definitely not canine. I have also seen a black cat (collie seized) in the area about 10 years ago and it was not that, it was a lot bigger.

Please let me know what you make of it and feel free to phone if needs be."

These pictures hit the headlines all over the world with the majority of people stating that the animal shown was a dog. Others pointed out how it looked like a different animal in each picture.

BIG CAT YEAR BOOK 2008

Some believe it was a shape-shifter, the place where it was snapped is Hound Tor, alleged to be haunted by phantom dogs, and the inspiration for Arthur Conan Doyles' *'Hound of the Baskervilles.'*

Martin Whiteley who saw the animal is adamant that it is a large cat. **(Source: *BCIB*)**.

18th June: aMail sent to Chris Moiser: *"Our wildlife crimes liaison officer reported a beast sighting they made, and said they would keep me informed. It was seen on 18th June at 6.45am at Bratton Flemming sports club. Described as single, large and black."* **(Source: *Chris Moiser*)**.

26th June: Sighting by four witnesses, including Ian Waldron at 7am. A black cat was spotted in a field adjacent to the B3227. Its tail was around 3ft long and its body was similar to a large dog, but low to the ground with the long tail. It was seen for at least 15 minutes at a distance of 400yds.

Mr. Waldron said: *"My wife and I were in bed and my wife spotted something strange walking on the edge of one of our fields. The field is planted with maize and has a 6m grass margin along the outside. The animal was walking along the edge of the grass margin, which would be about 2- ft high, stopping, and crouching. Our neighbour was in a field the other side of the road walking his cows in for milking about 50yds from the cat completely oblivious to what was there. It then got up and continued slowly like a cat to the corner of the field and disappeared out of sight. We had time to get two sets of binoculars to watch it and our two daughters (aged 7 and 9) had a long sighting of it.*

Incidentally my mother in law saw a glimpse of what she thought was a black cat about 3 weeks ago in a our five-year-old young woodland plantation which is in the next field to our sighting.

Unfortunately, it was too far away to get a meaningful picture but we have no doubt that it was a black leopard. All this we viewed from our bedroom window." **(Source: *BCIB*)**.

1st July: A sighting of a large black cat between Totnes and Paignton. **(Source: *BCIB*)**.

July: The new owners of the Dartmoor Wildlife Park reported seeing a 'wild' puma in the park! **(Source: *Christopher Johnston*)**.

24th July: *"As a result of a telephone call from Exmoor Zoo earlier today I attended a field near Bratton Fleming tonight and examined an 8-year-old horse and the field in which it had been present. The horse was a female, aged about 8, standing 15 hands high. It was and is a steady horse with a reasonable temperament. It was attacked between noon on the 23rd and noon on 24th. A local farmer reports his dogs barking excessively and abnormally during this period. The sheep in the next field were also huddled in one corner and frightened this morning. The injuries to the horse were 4 small scratches 20mm, 7mm and 15mm apart (i.e. total span 42mm), each 5 - 10 mm long; on the outer aspect of the right front upper leg. There are 2 puncture wounds on the inner aspect of this leg (i.e. "armpit"). These were deep and had extensive oedematous swelling around them to the point that it was not possible to measure the distance between them. There was also a scratch, which had bled badly on the chestnut below here. On the back left leg there was a V shaped scratch on the outer rear aspect of the upper leg and a 20 mm or so scratch on the joint below this. These when found this morning had bled badly. Inspection of the field showed it to be one of the best-fenced and orderly fields that I have ever seen. The fence was sheep wire with a strand of wire and then one of barbed wire above, it was all tight and well constructed. No refuse or rubbish whatsoever. None of the barbed wire was at the heights of the wounds. No other sharps of any sort could be found. The loose box in the field was again excellent and well maintained.*

There are no definite conclusions here, but no other explanation for the wounds than a large predator (but presumably smaller than an adult leopard or puma). The area is a regular beast of Exmoor sighting area, although nothing of note has been reported here for about 4 years." **(Source: *Chris Moiser*)**.

30th July: *"I am 44 yr old female, also an ex-trainee big cat keeper (from the old Sparkwell wildlife park) I keep two dogs and five domestic cats, so I am more than able to differentiate between animals of all types.*

On Monday evening, 30th July and evening of Wednesday 1st August I saw and watched for a fair while a very large, sleek black cat, (not a puma) with a long tail that was a little larger than my German shepherd. It was a very clear evening at approx 7.30 on the second it was approx 7.45. I was on the Dorset army camp at Blandford staying with my daughter and son-in-law, who is in the army. I'm quite a fit person and regularly went for a ramble on the moor right outside of the camp, alone, it is a fairly wooded area with an abundance of deer, foxes, badgers and rabbits.

I was on the first event climbing over a stile and saw the cat cross over the path I was about to take. It was about 25/30m in front of me, I was shocked to see it obviously but it paid no heed to me at all, it just carried on through a large hedgerow.

The second time I was sat having a drink of water and a cigarette when I saw what I'm assuming was the same cat, it shot out from bushes and in to a large open field that had, I think was recently cut hay growing, as it ran it was sort of springing until it stopped suddenly. I think it must have caught something as it went down out of sight in the hay or crop which was growing in the field, the following day I went to the same field and found a large area which was squashed down where the cat laid.

Unfortunately, I didn't take my mobile with me on these jogs but I so wish I did. I hope this helps with your research."
(Source: *Merrily Harpur* - www.dorsetbigcats.org).

9th August: A two-witness sighting at 11 pm in Devon/Cornwall. One of the witnesses, Sam Thompson, reported seeing a black cat on a dual carriageway between Exeter and Newquay (cut closer to Exeter). It was the same size as a Border collie, and looked just like a normal domestic cat but was a lot bigger at 3-5m, although this was seen whilst they were travelling at about 100mph. It was going to run out in front, but hesitated and stopped at side of road **(Source: *BCIB*).**

13th September: This morning on the local radio there was a story of a 'beast of Stoke Gifford' - the report was of a boy who had seen a creature on a path near his home in the Stoke Gifford area and it was hissing, was an albino colour with a 'face like a chinchilla' and 'red eyes'. **(Source: *BCIB -Elizabeth Andrews*).**

27th September: A large black cat-like creature was reported near Exeter Airport by a new Exeter University student on Thursday. Sheldon Rogers was driving to Exeter from Bournemouth when he spotted the animal at about 10.30am. He said: *"It was the size of a large Alsatian and it was just loping across the road in no particular hurry. It didn't seem bothered and just walked into the adjacent field after crossing the road."*
(Source: *Express & Echo*).

3rd October: We had some feedback this morning from a golfer who uses the course north of Newton Abbot. He told a chap that there has recently been a sighting of a large black panther-like cat close to Heathfield Woods and Trading Estate, heading towards the South West of Chudleigh? Apparently, the cat was crossing the road at around 5 am, and it was caught in someone's lights, the guy driving the car has sworn that it was definitely a leopard. **(Source: *BCIB*).**

Early October: A pile of animal bones discovered in a remote location, and the carcass of an Exmoor pony with one leg missing have prompted speculation that the 'Beast of Exmoor' could be on the prowl in the Lynton area. **(Source: *North Devon Journal*).**

2nd November: *"My neighbour three doors down saw a big black cat walking up his drive earlier this week (Station Road). It's not a very highly populated road - but even so - pretty amazing that it was walking up a driveway which would clearly smell of humans and the roadway is often walked. When the cat heard him move, it shot over into the fields behind. We are about ¾ mile from the woods.*

Today - another friend saw the cat again down in the fields behind Avon Mill Garden centre by the river. Again outside of the woodland. Why would it be leaving the safety of the woods during the day, is that usual? There is no livestock along here - only horses." **(Source: *Rick Minter*).**

3rd November: Sighting at 2am on Saturday reported by Hilary King, of Brayford, Thorn Park Cross of a shiny black animal, about the size of a Shetland pony, but not as fat. It was seen for about one minute, and it was about 20m away. Ms. King said: *"I was unable to sleep and went downstairs with the lights off. The security lights came on, which is not unusual, but when I looked out of the window the security lights over by the garage were on too. These only come on when there is something, or someone, lurking. I then saw this large animal making its way down the drive, it was not walking in a straight line, seemed to be walking from side to side."* **(Source: *BCIB*).**

8th November: A woman saw a large black cat disappear through a hedge - no location given. **(Source: *BBC Radio Devon*).**

10th November: A lady reported taking photographs of what she believed was a leopard, (left) to Anthony Bevan. She said: *"Me and my husband were 300yds from the cat, you could clearly see it up on the side of the valley and when it got closer they both said it was 20in high and 3ft long. A neighbour also reported seeing the animal on other occasions, he said: "it was not a leopard but a very big domestic type cat, again 18 - 20in high and 3ft long."*

Obviously this cat looks to be more than nothing but a domestic or feral, and after investigation, Anthony Bevan found the animal and confirmed that this was the case. **(Source: *Anthony Bevan*).**

11th November: A puma was spotted in the Chumleigh Valley. **(Source: *Anthony Bevan*).**

17th November: *"During this afternoon I have had 3 telephone calls relating to a sighting of a cat at Silverton, just North of Exeter. The first was from BBC Radio Devon and told me that a woman had seen a big cat, in her garden at about 2.30 on Saturday am. It had left through the hedge and she had got some hair from the spot. Would I be willing to look at it please? They could get back to me in a week.*

The second one was from the woman herself and she wondered if she could send me the hair (given my number by Paignton Zoo) She had seen the animal in good light and it was pale coloured - definitely not black. It was German shepherd sized and long and cat like. It moved in a cat-like way.

The third one was from the Western Morning News asking about the same sighting and looking at general big cat facts. The WMN is available on the "thisisdevon" website so I would imagine that story will appear in the next few days." **(Source: *Chris Moiser*).**

23rd November: Just had a call from a chap on Exmoor, who lives in a remote part, and the only access to his property is by a small lane. Basically, he told me of an ex-marine who found a deer half eaten in a tree - huntsmen had seen the cat (around in 83 marine hunt).

A large cat was coming up to female's bedroom, growling outside, and the, normally vicious, Alsatians were terrified. It has been seen four times over the last week, and is big and black. The female, whose bedroom it is, was in tears with fear. A huntsman also saw what he calls a black leopard attack, and take a down, a small deer in the last couple of weeks.

The Alsatians cower in fear, and will not go outside. **(Source: *BCIB*).**

27th November: *"I have some interesting photos from Exmoor taken over the* weekend. We did not see any big cats but found some very good evidence that we feel may be from 3 different big cats over Exmoor." **(Source: Christopher Johnston)**.

November: Sighting of a large cat-like animal near Hound Tor **(Source: BCIB)**.

December: Sighting of a large cat-like animal near Hound Tor **(Source: BCIB)**.

Dorset
12 reports

30th January: Report from Shirley Farrar. Seen at 14.30 in the woods at Littlemoor, Weymouth, Dorset. It was pure black with a long tail, carried low, and was seen for 20 seconds at a distance of 30ft. Ms. Farrar said: *"It was as big as my Doberman, I would say roughly 3ft tall, including the tail about 5/6 ft long. I was walking my 2 dogs through woods, when they flushed it out and gave chase. The dogs were going crazy and were very unsettled and edgy when they came back. I have also over the last 4 weeks found 3 deer carcasses completely stripped of all flesh, the last kill was only a day after the sighting."* **(Source: BCIB)**.

March: *"I thought I should E-mail about a recent sighting of a big cat in Netherbury. I was surprised, after reading some of the previous sightings, that our experience matched others in so many ways. I was driving from Bridport and took the first turning into Netherbury after Melplash. There were three of us in the car and just as we dropped down from the main road, we saw a very large black cat cross the road and walk through the hedge into the field to the right. It was about the size of a small Labrador but carried itself low to the ground. It wasn't in a hurry and just casually walked through the hedge. What really struck me was the size and length of its tail - very thick and heavy and seemed to hang down to the ground, close to the animal's body and then curl up again at the end. This encounter is so similar to some of the others - same place and same description, I just thought you'd be interested to know about it."* **(Source: Merrily Harpur -www.dorsetbigcats.org)**.

16th April: *"Today Jon McGowan, Mark North and I spent the day scouting round the Isle of Portland (Dorset) in quest of cat field sign. We got off to a slow start, but we ended up finding a ton of stuff.*

While it still seems difficult to appreciate how any cat might get on to the island in the first place (it's not an island but a peninsula, and to get on to the peninsula any terrestrial animal has to cross the long narrow isthmus connecting Portland to the mainland), many areas of the island look like ideal cat habitat.

There are quarries with plenty of cover and potential den sites, an enormous

number of caves and hiding places that are mostly inaccessible to humans, and a large rabbit population. Recent sightings have apparently been of lynx, and all the field sign we found was consistent with that identification. Our first discovery was of a definite cat track, hard-baked into the chalky mud on a pathway. It was of the morphology and size expected for a lynx and we found subsequent examples later in the day.

While passing a small rock-strewn gulley, I noticed the remains of a predated adult corvid (either carrion crow or jackdaw). We followed the gulley: a second predated adult corvid was at its head, as was scat deposited on the top of a rock. The droppings had the distinctive scent of cat scat and contained hair. We then found a great deal more scat - evidently cat in origin and not produced by dogs or foxes - along a ridge that was adjacent to a long line of enormous boulders and chunks of Portland stone. In amongst these rocks were a number of crevasses and alcoves. They were too tight for me to climb into but would be

ideal hiding places for felids. In one alcove I recovered the predated remains of a bird (probably a pigeon). Its sternum had been heavily chewed by a carnivore, but it was not possible to say whether a fox or cat had been responsible. One definite pile of cat scat was even deposited on a coastal path near a wooden bridge and would have been passed by tens of people every day. All the evidence we found was either collected or photographed. Sure, we didn't see any cats (excepting one pet moggie), but kills, tracks and droppings are the next best thing. Again, it brings home how easy - relatively speaking - it is to find such stuff if only people are prepared to look for it, and if they know what to look for. It was a neat day. We spent lots of time looking at lizards as well as seabirds and other wildlife. Thought you'd all be interested in this, all the best."

Darren Naish
School of Earth & Environmental Sciences
Burnaby Building, Burnaby Rd
University of Portsmouth
Portsmouth, UK, PO1 3QL

29th May: Big cat seen in a field to the inland side of the A39 between Kilve and West Quantoxhead. *"It was a large black non-domestic cat larger than a fox or badger and with a long tail. It was so obviously feline that we turned round and went back to watch it complete its walk from one side of a large field to the other - this was about 1p.m. and although the animal was 100yds away from us the field went uphill from the road so we got a good view of the shape and way the animal walked."*

Other details are as follows:

- Large dog size with a long tail - no binoculars but in sight for a minute - Black, couldn't make out any markings- Wild feline- Stealthily walking through the grass - Field with long grass, leading to hedgerows and trees- Driving - wife saw the same thing and we compared what we saw with our expectations of the size of a fox / badger, but it's movement reminded us of a cat. **(Source: *Merrily Harpur - www.dorsetbigcats.org*).**

28th August: Mark Dawson spotted a large black cat on the outskirts of Dorchester. He said: *"It was all black and had all four feet close together, the way a cat sometimes stands."* **(Source: *Dorset Echo*).**

4th September: *"I had a weird sighting this evening that scared the life (not the exact word used in the original report) out of me and I thought I would let you know about it. I will give as much info as possible, most of which probably won't be useful, but you never know and it may help you get a better feel for what happened.*

We live just south of Wimborne. Our garden is mostly woodland and as such we are very used to the wildlife that lives in and around the garden. As a rule I put out our dinner scraps every night along with some cheap dog food and whistle for the foxes, which nearly always come running. I say nearly always because for a week or so during the early breeding season they have other things on their mind.

Also when they kick out the cubs which happened earlier in August, they tend to spend a lot of time patrolling the boarders of their territory, which generally because of the layout of the road/woodland and the properties which have fences running east-west. Therefore, the foxes generally have east-west territory, as do the badgers and deer. Anyway, back to the sighting:

At approximately 21:55 (Monday, 3rd September 07) I had put the food out for the foxes, but after roughly 5 minutes of whistling, they hadn't appeared. This is very unusual as there are several foxes, which come from different directions, but this has happened a few times before.

On those occasions when this has happened it is either due to other fox priorities (mating, etc.) or sometimes when the weather is very bad. When it is very windy, or raining it can take them a while to appear, or they don't hear me or just stay under shelter and come out for the food later on (spoilt little buggers).

But very rarely and oddly around this time of year (if not a bit later in September/October) there has been a few successive nights where they have not appeared at all and the night is very still. In fact, eerily still. In addition, when this happens I get goose bumps and that 'being watched' feeling. This is mainly because there is usually always some noise from the rooks that nest in the trees, the squirrels and other wildlife. Usually something will make a noise when I whistle, or be making a noise anyway. As I said, when this happens, the foxes do not appear for a few days and each night has the same 'scary' feeling. Like a subconscious warning that, something dangerous is out and everything is keep-

ing its head down. Kind of a primal sixth sense type of feeling.

I must point out that I have no fear of the woods at night and quite often go for a walk on my own without lights and just sit and listen to the wildlife.

Back to the sighting. Shortly after coming in from the foxes there were a noise in the garden, which was loud enough to hear from inside the house. I thought it was a tree falling or large branch coming down. Which happens quite a lot, so I got a lantern and headband light/torch and went out to check for damage. I use the lantern for general light and the headband for a spotlight light on the trees.

I had a quick wander about and was just entering the edge of the woods next to my garage when something made me spin round to my left. I had not heard anything but something made me turn feeling almost startled, what I saw did startle me to say the least. Staring back at me from about 15m away was a pair of large green eyes. (I say green but I am colour blind and greenish is as close as I could say as it was very pale).

These were completely different from the usual fox's eyes that I constantly see reflecting my torchlight from the woods. Fox eyes are very round and quite close, and I would describe them as a silvery colour. These were also different to owl eyes, which are much closer and rounder.

These eyes were bigger and set much wider apart, at a guess from the distance I would say that they were nearly twice the spacing of a foxes eyes and from the height above the ground I would guess to be about 3ft up. At a push, I would be forced to say that the eyes were slightly turned down at the outside edge (if that makes sense).

Occasionally I accidentally sneak up on a deer and catch them with the torch, but they always bolt and you are lucky if you get a reflection from one eye. Even rarer I happen across a badger, but they always run off like the wind and they are generally very noisy.

This thing was silent, did not move and just stared me out. I could not see the body or head at all, just pitch black with the eyes staring back at me.

This in itself is also very unusual, as you can always see the pale fur of foxes cheeks or some light patches from other animals. Especially the white stripes on badgers and the light underbelly of deer.

I flicked the headband torch off to one side and back again, which can sometimes make a fox or other animal bolt off, but it did not move. So I thought 'Sh! T' and slowly side stepped back toward the house keeping the light on it until it was behind the garage. I must admit I did break into a slight 'Jesse Owens' as I came close to the door." **(Source: *Merrily Harpur - www.dorsetbigcats.org*)**

12th October: *"I have just arrived at work having taken the No31 bus service from Dorchester to Weymouth. Whilst I was on the bus just after passing the roundabout near the football stadium on the first field to the right (about 8:05am) I spotted a big cat! It was stood in the middle of the field, facing towards Dorchester about 200m away from the road. It was pure black, with a really long tail and I would say just bigger than an Alsatian dog. I had about 4-5 seconds to look at it and it was beautiful! Hope you can add this sighting to your history. I am aware that there was another sighting in the same area about a month ago. It was reported in the Dorset Echo."* **(Source: *Merrily Harpur - www.dorsetbigcats.org*)**

22nd October: Sighting at 10pm, on the main road from Wimborne to Cranborne, just before the turning to go to Verwood by four witnesses. The nearest colour they could describe it as is to liken it to a jungle cat. One of the witnesses said: *"I was driving home from an evening out with others when we saw what we first thought was a fox crossing the road. On looking again, we all said that is not a fox it's too big and the wrong colour. It did not stop, just carried on crossing the road - did not even look at the car. It seemed to be in a hurry.*

My older daughter said she saw the black cat a couple of years ago when she was returning home from a night out, but never reported it as we all teased her about it. That was along the same road further up near the crossroads one that goes towards Edmonsham (left) and the other that goes towards Sixpenny Handley (right) going towards Cranborne." **(Source: *BCIB*).**

1st November: A woman reported a big black cat in the Matchams Lane area at the end of runway 26, Bournemouth Airport. It was on the airport perimeter and definitely not a domestic cat, or fox, as it was too big. She was adamant it was a big cat. This happened at approximately 12.30am when she was driving home. I subsequently contacted security at the airport and was informed by the security guard that there have been numerous sightings recently, especially to the north of the airfield in the heath land within the airport grounds where not many people go. He himself told me that he had seen something whilst out on his security patrol recently. He seemed genuine enough. **(Source: *Merrily Harpur - www.dorsetbigcats.org*)**

11th November: At approximately 6pm the witness was on the road between Hinton St Mary between Marnhull and Sturminster Newton: *"I was driving towards Sturminster Newton and the animal ran across the road and jumped through the hedgerow just in front of my head lights. I hit the brakes but didn't get a clear sighting - just saw a long tail and long body and it moved in a cat like way*

My daughter (18) is also certain she saw a big black cat in a field about 6 weeks ago on the same stretch of road - we turned the car round but it was gone when we went back.

Very long tail which was the main thing I noticed." **(Source: BCIB)**.

6th December: Sighting at 8.15pm at Wool : *"I was driving along a small road near West Holme and saw white/blue eyes in the headlights on the left side. I stopped about 6ft from the creature and had a clear look at it for a couple of minutes as it stared at the car lights. I was stunned to see it was a panther - I had to pinch myself. It had something in its mouth, possibly a squirrel. It was the height of a Labrador dog and about 4-5 ft long not including the tail which was thin and long. It had a very angular face with whiskers, and a strong, muscular body. It seemed as mesmerised as I was, but when I made a movement to lock the door (irrational I know) it backed off slowly into the woods still staring at the car lights."* **(Source: BCIB)**.

Mid-December: Sighting during the late afternoon mid Dececember 2007 at West Knighton quarries near Dorchester. The animal was completely black, with a thick tail approximately 2ft in length. Its height was 18in approximately, and the body - not including the tail – was approximately 2ft. The witness said: *"I was walking my dog. The cat crossed the track about 15 to 20m in front of me. There was no reaction from my dog, possibly because he didn't see it."* **(Source: BCIB)**.

Essex
14 reports

25th January: Sighting at 8.45 am: The witness reports: *"I was driving on the M11 and was 1-2 miles North of the Harlow turn off heading towards Cambridge when a big cat, sandy in colour with dark patches, ran across the six lanes ignoring all traffic."*

The cat was impervious to the oncoming traffic, and sauntered across the six carriages of the M11 glancing for a moment at the oncoming vans and only making it across to the other side by a cat's whisker. I've phoned the Essex police to ask if there were any zoos in the local area missing a lion cub. A few years back I saw a large black puma-like cat in a field in Little Christmas near Ware in Hertfordshire (I think it's called Little Christmas--it was down a country road)."

He further describes the cat as having a long tail held up in a "cat-like fashion." Front paws "dragged forward. He states it looked like a lion cub. **(Source: BCIB)**.

March 2007: Patrick Griggs, who lives next to Toot Hill Golf Club, believes droppings in his garden and an unusual cat-type noise could prove that the so-called 'Beast of Ongar' continues to be roaming the area. **(Source: *Epping Guardian*)**.

1st April: A black cat, really thin and skinny spotted several times over the last three days. Slightly bigger than a Labrador with an 'S' shaped tail. **(Source: *BCIB*).**

4th May: Sighting at 09.40 at Halstead, Town Centre, just off Morley Road. It was black, although it did appear to have a patch on its left side. It had pointed ears, a long tail, and was 25-30ft away. It was hard to judge its size, but it was considerably bigger than a domestic cat.

The witness reported: *"There is a deserted spinney at the end of our garden which is close to allotments and a graveyard, the public do not have access at all, muntjack deer also live here and we often see them. I let my dog into our garden and he was excited by something through the chain link fence. When I looked I could see approx 25ft away two large pointed ears, which I took to belong to a fox, as they were so large and pointed and facing forward. The animal was sitting in amongst some low undergrowth between some small trees in a patch of sunlight. I watched for approximately 4 minutes and could see the ears rotating and the animal was clearly watching me. Thinking it might be an injured fox I jumped over the fence to see what it was at 10ft away the animal was still watching me, as I moved closer it jumped up from its laying position and ran to the right. It was a very dark colour with very long and thick tail large body and head and distinct ears. It appeared to have a lighter patch on the side of its body but that could have been sunlight. And most definitely bigger than a domestic cat. I called my wife on the phone and told her she also said that she has seen the same cat in recent weeks but didn't tell me for fear of ridicule."* **(Source: *BCIB*).**

10th May: The 'Beast of Ongar' has been spotted in fields near Abridge and spooked a dog walker and his dogs in Toot Hill. **(Source: *Epping Gaurdian*).**

23rd May: Remains of a young deer found on Mount Road near Theydon, clawed from the back, some believe it is the victim of a big cat. **(Source: *Epping Gaurdian*).**

14th July: Seen at Staple Field House, North End, Halstead, Sudbry.

The witness said: *"The animal jumped and hit my exhaust on my motorbike. I was riding my motorbike when the black animal figure jumped at me it hit my exhaust. I think the animal fell to the ground because there is a lay mark in the corn field."* **(Source: *BCIB*).**

17th July: A pilot from *Ryan Air* has reported to Paul Westwood's Big Cat Monitor's website that he spotted a black panther-like animal while making his way home from Stanstead Airport. It was at 00.15hrs on Butlers Lane between Saffron Waldron and Ashdon. **(Source: *Paul Westwood*).**

30th July: Monique Jowers saw a large orange cat with red spots attempting to enter her house through the French windows. **(Source: *Enfield Independent*).**

31st August: Seen by Mark Malyon at 3.55pm, in Brentwood, Essex on Dark Lane. The cat was all black, with a 2ft long tail (or more) and was 2 or 3ft high and 4ft long (without tail). It was seen for 1 minute or more, at a distance of 600 -700yds, whilst the witness was sitting in a car down a dark lane - the cat was sneaking down the hedgerow. **(Source: *BCIB*).**

15th September: Sighting of a black cat by Roger Hart at 17.30, spotted along Western Avenue, Epping. The tail was large - ½ to ¾ length of body, and it had rounded ears. The animal was 18 in height 28-30 in length and the duration of sighting was 6-10 sec at 80-90 ft. Mr. Hart said: *"I was walking my dog down the lane from Western Avenue to the common when the cat crossed the path from one side to the other in front of me."* **(Source: *BCIB*).**

7th October: I don't know if this is of interest to you but there has been a sighting of a large black cat in Colchester (by North Station). The old chap who swears that he saw it was sober and walking his dog at the time. **(Source: *BCIB*).**

7th November: Seen at 17.55hrs at Newport, Bury Water Lane. The animal had black, rounded ears, and was seen from about 75-100yds away. The witness was not too sure about the size as he only glimpsed the animal in the light from his lamp, but it was maybe about the size of a Great Dane. The witness said: *"I was walking my dogs, and the cat just stopped, then disappeared into a small copse. I have seen a very similar, if not the same, animal in the same area. I think it is a black panther that I have seen and heard. Too many normal people have seen similar animals, but are too afraid to go public for fear of being made to look a fool. I am not sure what to do about them as they do not appear to be doing any damage."* **(Source: *BCIB*).**

13th November: Sighting at 11.30pm in woodland area off Stansted Airport by two witnesses. It was jet-black with green eyes, and had rounded ears. Its height was about 1m. On of the witnesses said: *"I was driving home when the cat ran into bushes one metre away from me."* **(Source: *BCIB*).**

Gloucestershire
31 reports

23rd January: On January 23, a 38-year-old driver from Lydney was heading along Valley Road towards the Bridge Inn at 9.10pm when he saw a black panther-like animal crossing the road. It did not look up and continued to amble into the darkness of neighbouring Linear Park. (***Gloucestershire Echo***).
6th February: Firefighter Peter Bishop says he came face-to-face with a black

panther as he walked home along Causeway Road, Cinderford. He froze to the spot but the big cat just looked at him and strolled off. **(Source: *Gloucestershire Echo*).**

February: A big cat killed two sheep at Hartpury, with bites into skull (no photos). Farmer's friend, who is well informed on cat species, later watched it with spotlight and said it looked like a leopard-puma cross.
(Source: *Frank Turnbridge*).

14th February: Sighting by Dower House along the M32. Black, very big and very long in a stalking position, with a very long tail.

The witness said: *"I was driving towards town on the M32 at east at 0735, when I saw a very big black cat, the size of a puma. I told all my family about it and researched black cat sightings that night to see if any had been spotted in the area. I didn't report it as I didn't think that I would be believed. Then, when I got home tonight, my Dad told me about the news item on HTV local news re: black cat sightings in Stroud. How strange!*

It confirms to me that what I thought I saw was actually a big cat. Pretty worrying - where can it go from the M32. (The cat was by the big old former hospital; I think it's called Dower House) " **(Source: *BCIB*).**

14th February: *"I was standing by the fence at the top of Football field in Upper Siddington and looked down towards the twenties (houses at the bottom of the field) and saw a large black cat with a long tail, about the size of a springer spaniel, crossing the far end of the field. Iit was in no hurry with a definitely cat type walk, and it disappeared into the undergrowth by the canal. This happened at about 11am."* **(Source: *BCIB*).**

19th February: - see 14th – *"Today, the 19th February, my brother saw the cat in the same field at 6pm but this time on the other side of the field walking by the fence by the road nearer the top of the field. He also said it was black with a long tail, but he was closer than I to the cat."* **(Source: *BCIB*).**

19th February: A security guard told police he saw a large black cat that he thought was a panther. Gloucestershire Police said it happened at 1.30am on Thursday at Wormington Gas Station, between Tewkesbury and Evesham. **(Source: *Gloucestershire Echo*).**

Mid February: Police believe a pair of big puma type cats are prowling in woodlands near Stroud after the remains of three deer were found in the last two weeks. **(Source: *Stroud News & Journal*).**

25th February: Sighting at 5:30pm at Down Ampney of a large and black cat with rounded ears, and a very long tail. Its length was about 6ft long, including the tail. The witness said: *"I was playing golf in a football field when the ball went into a corn field. Whilst looking for the ball there was a large growl and then about 3m from me it made three large dashes over the distance of about 20yds and ran into the forested area nearby."* **(Source: BCIB).**

21st March: At his car boot on Saturday, Frank Tunbridge was told of another sighting on edge of Randwick woods last week - the chap was a sceptic but was 20yds from a collie size black cat and his Lurcher, who chases cats and will kill domestics given the chance, was straining at the leash. The chap returned from the car with his mobile phone camera 10 minutes later and saw it going the other way, but too far off for a photo. **(Source: *Rick Minter*).**

3rd May: My talk last week in Stroud has prompted this sighting report from a lady in Selsley, near Stroud. Frank Tunbridge and I have had witness reports here in the recent past and close by at Randwick & Standish woods, which is in same territory. It is clearly a live area for 10 years.

"I saw a large black cat on 3rd May between 7.15 & 7.30 pm in a field adjacent to where we live at Water Lane, Selsley East. A couple of days later I was chatting with our neighbours and they said that their son (who is in his early 20s) was followed by a large black cat as he was walking up Water Lane. I don't know at what time of day this was, or how long ago, but when he told his friends, he was laughed at and apparently hasn't mentioned it since. I reported my sighting to DEFRA Wildlife Division & Glos Wildlife Trust, but no one seemed very interested." **(Source: *Rick Minter*).**

2nd June: (Rick Minter reports) Frank Tunbridge and I are still on the case with the jungle cat(s) at Port Ham, edge of Gloucester. Frank has heard it twice in last three weeks (identical to the calls of jungle cat on YouTube) and got a glimpse of a grey, fox size creature running in the dusk last night - I was around the corner and missed it. We've set up stealth camera but one was nicked and another flooded. **(Source: *BCIB*).**

11th June: Sighting by James Sleep and two witnesses in Bredon, Cheltenham. The weather was pretty warm, perhaps a little muggy. The cat was black/charcoal with a 2ft tail and black, short hair. The witness describes the cat as a jaguar, and was seen from 50 m away for around twenty seconds. It was described as being 2ft in height, but the witness could not determine the length of the animal. The witness was shooting rabbits at the time when he spotted the cat. **(Source: *BCIB*).**

June: 13 year-old Jody Motterham told his parents he'd seen a black panther in the woods they didn't really believe him. **(Source: *The Forrester*).**

14th June: Milkman Robert Brinton could hear what he thought were two garden moggies fighting as he delivered in Railway Road, Ruspidge. But when

walked round the corner he saw one of them was more than the average puss. *"I could hear cats meowing and growling,"* he said. *"This thing had a pet cat backed up in the grass."* Robert said the beast, which ran off into nearby woods when it saw him, was black with short fur.
(Source: *Gloucestershire Echo*).

14th June: An off duty special police constable spotted a big cat the size of a Doberman heading from Railway Road towards the Dilke Memorial Hospital, Ruspidge. Two hours later a postman spotted a similar animal in the exact same spot. **(Source: *Gloucestershire Echo*).**

10th July: An American couple contacted the Observer after spotting what they believed to be a large cat in fields outside Chipping Campden yesterday. Environmental consultant Rick Minter said: *"The Cotswolds is a perfect setting for these animals, with plenty of deer, extensive tracks, woodlands and valleys."* The couple, from Mississippi, said they were flabbergasted and couldn't decide if the cat was a lynx or something bigger. **(Source: *Cotswold Observer*).**

July: Sighting in the Forest of Dean. A lady was driving at Park End (area of several sightings over recent years) when a deer ran across the road. As she put brakes on she saw a large animal with a scruffy mane in hot pursuit of the deer. From colour and size, it seems to be closest to a puma description. **(Source: *Rick Minter*).**

30th July: Seen at 21.20hrs, in fields near Lydney in village of Whitecroft.

The witness said: "I never saw anything, but heard distinctive groan/guttural growl from thick undergrowth whilst hunting with my rifle. I think whatever creature was present was warning me that if I got closer, look out! I am 39 and have a very keen interest in nature and I am sure what was heard was feline. I have good knowledge of indigenous mammal sounds including wild boar, and this was definitely alien to my hearing, except for what I have heard on TV - it sounded like a large feline.

The growl came from the same field as 30x calf, (approximately 150yds away). I counted all 30 over the other side of the field. I will return tomorrow to seek further evidence, with shotgun for company. I was approx 15/20yrds away.

I believe an accident is imminent. If I had walked into the undergrowth from whence the sound came and happened across a large feline, I don't think I would be writing this sentence. I think they should be caught." **(Source: *BCIB*).**

6th August: Frank Tunbridge got a call from a deer stalker near Sudeley in Gloucestershire (where there was a sighting last year of a big black one). The stalker saw a big black cat, plus two cubs. **(Source: *Rick Minter*).**

27th August: *"My family and I were out walking in the fields behind Stratton, Cirencester today at approx 3.30pm when we saw a large black animal in the wheat. My wife saw it first, about 100m distance. I didn't really pay any attention right away as I assumed it was a black Labrador.*

I diverted my attention to looking for its owner in case it was someone I knew. But there was no-one else around. The animal was taller than the corn and appeared to be standing looking at something on the ground - we could not see its head or its tail but if it had been a dog, we would have seen the tail. The black hair was smooth. It did not scamper about like a dog, it was slow and 'stealthy'.

Then it went to ground and disappeared from sight. We stayed to watch for a while and then moved on. We did not see it again. I couldn't investigate further as the kids are young. Our dog, a Tibetan terrier did not seem unduly concerned, but this is not unusual - she does not show any hunting instinct.

I attach a map of the area with the approximate location marked. The wheat fields to the north & North West of this location were all cut short.

I have lived in the country most of my life and walk in these fields at least 3 times a week. I have never spotted anything like this before but regularly see deer, hares, rabbits, foxes and other animals. I would describe myself as being knowledgeable about wildlife and unlikely to mistake a deer or some other mammal for this." **(Source: *BCIB*)**

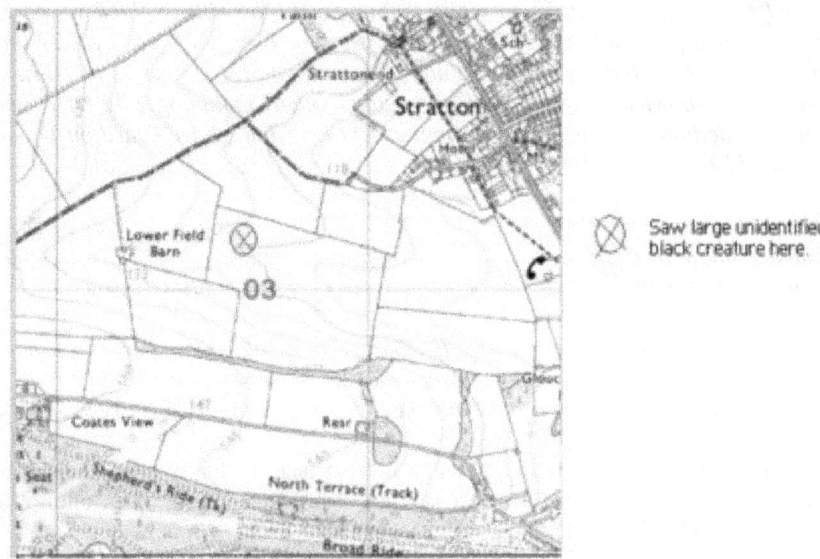

7th September: *"My daughter has just got home to say her friend's dad, a farm worker, witnessed a big black cat in the next village towards Tewkesbury, Prior's Norton, on Friday evening, 7th September. The chap's daughter B was in the car and saw it too. I was only two minutes in front of the car when they had the sighting, as I was delivering my daughter's friend (the witness's daughter A) back to the house. I passed the chap's car on my way back, but he didn't realise it was me to tell me what he'd just seen! Alas I have busy two days and can't break free to follow this up."* **(Source: *Rick Minter*).**

17th September: Sighting in Down Ampney area of Glos (north of Cotswold Water park, and fairly dependable for sightings over the years)... w/b 17th September. A farmer and wife observed a lynx from kitchen window around 9am. They thought it might have been ratting outside the barn. **(Source: *Rick Minter*).**

Late September: Again, through the grapevine, I've had a report of a couple walking their dog on the approach road-drive to Wallsworth Hall, Nature in Art. A big black cat ran across the road near them, and both were shy about admitting to the other that it appeared to be a big cat. This was two weeks ago and Frank Tunbridge is contacting the witnesses for more info. It's 2 miles from two previous sightings in Norton in early September, where I now have my remote camera up on the farmer's land close to where he watched the cat. **(Source: *Rick Minter*).**

Early October: A big cat sighting near Dymock by Forest by college student 16 year-old Natasha Baker. **(Source: *Gloucestershire Echo*).**

2nd November: Sighting at 10am on bright sunny day, with excellent visibility. The witness said: *"As I was driving on the 'back' road between Andoversford and Winchcombe, just after the right turn to Charlton Abbotts, I saw a cat like animal half in the roadside hedge on my left. It was about the size of a large dog but with a more slender build, fawn colour with dark brown spots and a medium length tail rounded at the tip. Most similar cat found in reference is a Lynx. Unfortunately, I didn't see its head. As I drove past it disappeared into the field beyond."* **(Source: *Rick Minter*).**

5th November: Two witnesses had sighting at 8:30ish pm, on the Ebrington/Ilmington Road. (Ebrington side of the old tip just past a copse on the left side of the road).

One of the witnesses said: *"It was so quick, and such a shock, but at a guess maybe 50cm long maybe 2-4cm thick jet black. We did not see its head. I saw its body, which I guess, would be 50 to 60cm long. As it was going through the hedge and downhill I wasn't able to judge its height.*

My daughter was driving me home after a bonfire and fireworks at Blackwell village hall. We drove over Ilmington Hill and as we came around the corner and passed the old tip in the full headlights I saw a huge black cat climbing down into the hedge. I said oh my god and Jennie said did you see that? We were so shocked and hardly believed our own eyes. We had not had anything to drink!

My husband and I were going to go and see if there were any paw prints in the field, but haven't had time yet. In the past I have seen in the papers about sightings and my brother-in-law has seen what he thinks were paw prints in his fields, but I would have thought there would be more sightings if there really were large cats out there. My daughter said on Monday 'Had I not really believed there were big cats around?' Well now we know there are (but convincing the family is impossible).

I think as long as they are no danger to anyone they could be left alone." **(Source: BCIB).**

19th November: *"I've just been phoned by John Beart. He himself has had three sightings in Glos. He was phoned today by a friend who was driving at Whiteshill, Glos, at mid afternoon. She had to slow down for two big black cats crossing the road (don't know exact size – see below re more info). The location is close to the reliable Randwick & Standish woods again. The lady was not that surprised as she'd been to Frank and my talk in Stroud in May. John will get fuller details and will report back."* **(Source: Rick Minter & John Beart).**

Approx 1st December: At his car boot sale on Sunday, Frank Tunbridge got told of another black panther sighting at Whiteshill-Randwick, two weeks ago. The cat was walking across the road in front of the driver. The driver is known to Frank and has had three previous sightings in Glos. This latest sighting was just a few hundred yards away from the location where two big black cats were seen crossing a road on 19th Nov. Frank will be putting up a camera at the valley bottom of this area, which we believe is part of the route down from Standish-Randwick woods. This area remains consistent for sightings, virtually month on month. **(Source: Rick Minter).**

10th December: *"I quite often travel by train for my work, and have increasingly stopped reading so much and started cat watching out the widow instead, given that it's such a good opportunity to have a good look across the landscape.*

I normally see deer and the odd fox that I otherwise wouldn't have noticed. This morning was a clear dawn on the way to London, and on a woodland edge a few fields before Didcot power station, at 8.15am, what did I see... a classic black outline of the panther (as if Mark had put his scale cutout there). My only reservation was that it was too perfect, and it wasn't moving within the say 7 seconds I watched it. But it was the correct shape, scale, proportions and jizz.

Just after Didcot at a similar distance (say 200yds), someone was playing with a collie dog, which was distinctly different and made me feel more confident about what I'd seen. So, I'm mildly excited, and feel the time I miss reading on the train is now worth it. Reminds me that Jonathan McGowan says one of his best sightings (a black leopard he thinks) was from a train.

My one good sighting previously (Shap Wells, Cumbria in 98) was a bit closer (c.120yds) but I was able to watch it walking for 40 seconds, and the movement helped verify it in my own mind." **(Source: *Rick Minter*).**

27th December: Message just in from one of my ecologist pals in Glos.

"One of my vols (mink trapper and ex deer stalker) was driving along the A417 between Lechlade and Fairford, and drove past some more unusual road kill. Near to the petrol station near the Whelford turning.

It was a mammal (!!), same size maybe slightly larger than a spaniel, black and glossy in colour. He dismissed it as a Labrador etc. but felt the pelt was too dark and glossy... He went back to check the next day for me and the carcass had gone. Sorry this is slightly vague..." **(Source: *Rick Minter*).**

Hampshire
6 reports

30th March: *"My husband found prints whilst walking our dog on the Perham Down Rifle Ranges in Perham Down near Andover in Hampshire."* **(Source: *BCIB*).**

May: *"I have just arrived home after spending a day on Hayling Island. We were travelling northbound on West Lane, when I saw the unmistakable shape of a large cat crossing the road ahead of me. The animal was honey-yellow colour, and crossed the road relatively slowly, however, when we got to the place were it crossed I could not see any trace. I had to look big cat sightings up on the internet because although I had heard rumours, I presumed they were just urban myths."* **(Source: *BCIB*).**

12th June: Joe Lea emailed the *Hayling Island Today* and reported: *"On June 12, I saw what I thought was a large fox. It was light sandy coloured, with a bushy tail that touched the ground as it walked. The animal stopped and looked at me, without any apparent fear. You can imagine my surprise when I saw that it wasn't a fix but a large cat! I was about 15ft away and I could not have made a mistake. It definitely had a cat's head with pointed ears and a flat face. The animal was twice the height of a domestic cat."*
(Source: *Hayling Island Today*).

26th July: Sighting at 10.30am near *The Fighting Cock* public house, Godshill, New Forest by four witnesses. The colour was pure black, with the tail down to the animal's knee part, with a curl at the end. The size of the animal was about Labrador size, maybe a little bigger, and one of the witnesses believes she saw a black leopard. She said: *"The cat was behind a car that was parked up and as we drove past, we must of scared it and it ran off towards the trees. If they are posing no immediate danger then I think that they should be left to roam freely. Maybe tagged and monitored. No ideas on where they originate from unless released from circuses."* **(Source: *BCIB*).**

12th September: Seen in Hambledon by two witnesses: *"Brownish/gold in colour, with pointed ears, and a 1m long tail, curled at the end, with dark patches. I would judge it to be about 1m in height, maybe more, and around 2 - 2.5m in length. It was difficult to judge from the distance. It was a large unusual feral. I think this is exciting for British wildlife but may be bad for the eco system, and if they got into towns or streets they could cause serious damage."*

(We would love to hear your comments on this picture: write to us at 35 South Dean Road, Kilmarnock, KA3 7RG, Ayrshire, Scotland.) **- (Source: *BCIB*).**

Early November: A big black cat described as definitely not a pet' has been spotted in Overton's Sapley Lane Skate Park, reports the *Andover Advertiser*. A woman reported seeing a three-foot-long black cat jump over her garden fence. **(Source: *Andover Advertiser*).**

Hertfordshire
9 reports

28th January: Seen at 2.30 pm, by the side of the A505 between Baldock and Royston, approx 3 or 4 miles southwest of Royston Dry, on a mild, quite dull and overcast afternoon: The witness said: *"It was a mottled sandy / brownish colour, a bit like a very large tabby. It had pointed / tufted ears, but I couldn't see the tail. I was probably about 100ft away when I first noticed it and about 20 or 30ft away as I drove past, and only glimpsed it for a matter of seconds.*

I was driving up the A505 towards Royston on a fairly isolated stretch of road, when I saw the cat on the grass verge, between the hedge and the road itself. It was completely still as if it had its eye on some prey and was waiting for the right moment to pounce.

The cat stood out to me primarily because of its size. It was the size of a small lynx or medium sized dog, but noticeably bigger than a domestic cat (even a large one). Given that I was some distance away when I first saw it and thought about its size, it would have appeared even larger close up. It was mottled sandy / brown colour, almost like a sandy / brown coloured tabby, but without distinctive spots or stripes. I do remember noticing it had distinctly wild-type facial markings, not like a domestic cat at all. I think its ears were fairly pointed or tufted, maybe both.

Maybe it was the shape of the ears that contributed to its wild-type facial appearance. I didn't see its tail. The cat didn't move as I sped past at about 70 mph. I wanted to turn around and have another look but there was no easy way of doing a U-turn and going back, so I drove on home to Royston." (**Source: *BCIB***).

25th March: Terry Goulden, who makes props for Hollywood movies, is convinced he saw a black leopard hunting just feet away from him on his land by the New River in Ware. (**Source: *Hertfordshire Mercury***).

13th April: Jon Harvey, 30, of Greenways in Hertford, spotted what he believes was a big cat while out jogging in Stapleford at 8.15am. (**Source: *Essex & Herts. News***).

21st April: Police armed with shotguns and rifles hunted a big cat believed to be on the loose in Hertford. Two officers spotted a "large, feline-type animal" on the prowl by the police station in Stanstead Road just before 7am on Saturday. It sparked a three-hour search involving 21 officers, including two firearms units, and the head keeper at Paradise Wildlife Park in Broxbourne, Colin Elcombe. (**Source: *Hertfordshire Mercury***).

Late May: 'Charlotte Church savaged to death in the Beckham's Garden' screamed the headlines.' Charlotte turned out to be a sheep kept by telly chef Gordon Ramsey for the table in his latest TV Show, the F-Word. The Beckham's gave permission for his sheep to graze on their land (Sawbridgeworth).

The sheep was apparently savaged and killed by a large cat that has often been reported in the area. When a vet was called to analyse the carcass, he told Ramsay and the film crew he suspected the lamb had been killed "by a big cat". **(Source: *Daily Mail - Hertfordshire Mercury*).**

25th July: Seen in St. Albans, (by King Harry Lane, nr. Waitrose). The cat was beige with black stripes, 3 ft high, 500 cm long, and had tufted ears. Its tail was 75cm long.

The witness said: *"I was sitting in the car watching it and the jaguar was just sitting on the top of the bottle bank while I put some bottles in another bottle bank."*

The witness also says they saw a panther near Batchwood. **(Source: *BCIB*).**

7th October: A gentleman has just called me from Hertfordshire, he saw a big black cat on Sunday morning, it was just off the M25 near to the A41 to Tring. He would like to meet someone at the location if possible. If there is anyone in the area and wants to meet him if you could let me know that would be great. **(Source: *Chris Johnston*).**

5th / 8th November: Black cat seen, near Croxley Green, Watford, Herts and Great Gaddesden, Herts (near the Ashridge Estate) which is reported to have released some back in 1976. Its tail was very long, and was s-shaped and very prominent. One sighting was about 5ft away, the others from a far greater distance.

The witness said: *"My friend spotted a very large cat, chocolate brown in colour with mottled black markings on Tuesday 5th November 2007 at about 8.00am whilst taking her children to school. The sighting was in Chandlers Cross on the way to Croxley Green near Watford. It seemed to jump into the hedgerow, which is when it caught her eye. She knew she was looking at a large animal at the edge of the road and it seemed to realise lots of cars were there, because she just saw it turn, very graceful as a cat would. It was about 3ft tall, and what she saw as the animal turned was a very long muscular body, but it was the length of the tail that stood out. A very long tail. She reported the sighting to the local police who took her story very seriously, needing the exact location as to where the animal was seen.*

When she told me her story, it reminded me of my husband's account about four years earlier. We have a farm in North West Hertfordshire and one day he noticed a large black cat running through the fields, we thought about a newspaper story we had recently read about a black panther in Wales seen attacking an animal, we laughed about his sighting and decided not to think about it anymore as people would think we were mad and also we did not want to alarm the children.

The family were out for dinner the evening after my friend's sighting and my niece mentioned my friend's story. One of my sons told us about a sighting he had had... I was shocked, although he had told people he never mentioned it to me. It was about two years ago, this sighting was also at our farm. A fellow carpenter and himself were working on fences, when my son heard a roar. As he looked up he saw two large black cats chasing three deer across the top of the fields, this side of a wooded area, they then ran into the woods and disappeared. Last night I checked with his workmate, who confirmed he saw a large cat like creature chasing some deer, it happened so quickly they were gone in a split second. He does not remember a roar although my son reckons it was the roar that made them look up.

Telling my other niece today about this letter I am writing to the newspaper, she informed me that about four years ago, we were having a small family barbeque at the farm, her small dog and herself were having a walk through the fields, when she spotted the back of a large black animal with a very long tail. It was walking along the same path that my son and his workmate mentioned.

The animal then crossed into the woods also. She had heard stories of big black cat sightings in Wales just before so she just thought she was being silly, she convinced herself at the time it must have been a deer. She had just moved out from London and joked today that they only had giant rats to deal with there.

Now I know this seems like too many stories all at once to seem credible, but the timescale is over four years.

After my friend's story, the others have just come to my attention and I am finding the whole thing fascinating. I would like to find out from your readers about any other sightings they may have seen or heard about. I have been told a number of them were released into the wild around 1976 when strict rules concerning the keeping of dangerous animals were introduced.

Are these animals territorial? Do they tend to stay in their own neck of the woods or do they travel all over the rest of the British Isles in search of food? Do they have a taste for deer? We know there are plenty around here, I ask this because we have not lost any livestock nor have any of our neighbours as far as we know." **(Source: *BCIB*).**

Kent
18 reports

1st January: A woman spotted a large black cat in the Chequers Road area of Minster, on Sheppey. (**Source:** *Sheerness Times Guardian*).

14th January: A grey to light brown cat was spotted on the old Canterbury Road, Ashford at 04.30hrs. There was no tail noticed on the animal, which stood around twenty in to two ft tall and around two and a half ft long. The witness reports: *"I was sitting in the passenger seat of a car. The cat was loping or moving in small jumps as if it was getting away from the car after crossing the road or maybe it turned around to get away from the car. The cat appeared to me to be like a small cheetah, especially the way it loped off, but I cannot remember seeing a tail."* (**Source:** *BCIB*).

Approx 2nd March: A large mysterious cat blamed for scaring horses in a paddock in Cold Blow Crescent, Bexley. (**Source:** *News Shopper*).

17th March: Sighting at Bexley village in St Mary's church graveyard. It was predominantly grey, with mottled black patches, and pointed, tufted ears. The tail was around a foot in length, bushy, grey with same colouration as rest of body. The large unusual feral was seen for about 30 seconds - 1 minute, at a distance of 15m. It was around 1 -1½ft high, and 2-3ft long, with large paws, around the size of an adult border collie.

The witness said: *"I was cycling along the footpath alongside the cemetery going from Bexley village to Joydens Wood, when the animal appeared from the area close to the cemetery rear entrance.*

Originally, I mistook it for a dog, and slowed almost to a standstill (I have been chased and snapped at by dogs when cycling fast) waiting for the owner to emerge. I then noticed the animal had pointed, tufted ears, and the nose/muzzle was short and rounded, with clearly visible whiskers, also the paws were larger than seemed normal. I stopped and took out my phone to take a photo, and my phone rang, scaring the animal off. It stopped dead when it came into view and stood and stared at me until my phone rang, at which point it crawled under a hedgerow and disappeared into the cemetery." (**Source:** *BCIB*).

18th March: A couple walking in a field beyond the Kent & East Sussex Railway reported seeing a large black cat on Saturday afternoon. The witness said: *"It was the size of a collie dog and bolted from the undergrowth, 30ft away."* (**Source:** *Kentish Express Ashford & District*).

25th March: A man spotted a large cat-like creature in Ashford and it may have been responsible for killing a family pet. He saw it from his home on Hamstreet near the railway at 6pm. He said the animal was as large as an Alsatian and was walking along the Hastings line towards the Marsh. The animal had a long black tail. (**Source:** *Kentish Express Ashford & District*).

2nd April: A woman has described how her "blood ran cold" after she heard and saw evidence of what she believes is an escaped black panther. The woman, who does not want to be named, heard the noise of an animal growling in the back garden of her home in The Grove, Biggin Hill, around two weeks ago. She was in her garden this morning when she noticed deep scratch marks reaching almost 20ft up the trees. (**Source:** *News Shopper*).

April: A large black cat was spotted in a paddock full of horses near Cold Blow Crescent. A resident saw the animal in fields at the back of his house, and noticed the horses were "very spooked." (**Source:** *News Shopper*).

21st April: Betty Morris spotted what she believes was the 'Beast of Bexley' stalking the streets around Becton Place, Northumberland Heath. (**Source:** *News Shopper*).

1st May: A woman spotted a cat-like animal the size of a Labrador in woodland near Chislehurst. (**Source:** *News Shopper*).

Early May: (Welling area) *"I have just been called by police contact regarding a man who lives near me. The past two mornings at 3am, he has watched a black big cat in his garden. The neighbour's security light is triggered so he gets a good view. This man has lived in West Africa, and says the cat is all black, smaller than a male (he's talking lions) and says it has a swollen belly. He said he thought it was a pregnant female.*

Police checks show he is who he claims to be. He doesn't want press involved. Nor anyone else. Only speaking to me because police deal with me. An older gentleman, seems solid. Spoken on phone, but am hoping to visit his house. Lot of activity in this area again, seems May to June it's at its height. Always thought mine was male, if this is all true and it's a female he has seen, we may have two, which would explain some size differences in sightings. Will give you further info as I get it." (**Source:** *Nicole Webb BCIB*).

May: Mr Carter, spotted a huge cat running along the dyke as he drove a van on Eastbridge Road, Dymchurch, at around 10.15am .He described it as black in colour with a rusty-red tinge. (**Source:** *Kentish Express*).

Approx 9th June: Gary Simpson travelled to work on his bike down Frith Road at Aldington when he spotted a large black animal sat in the road near the *Good Intent* Pub. (**Source:** *Kentish Express*).

24th June: *"Do the big black cats attack horses?*

The reason for my question is that my daughter's horse was attacked last weekend by a rather nasty animal - we have no idea what. She was mauled on the front and back left legs, quite badly. Minimal injuries were sustained on the right leg. I have taken some photographs for you to look at. We have found prints, and also scratches on a nearby tree. Our vet said it was definitely an animal attack but unlikely to have been a dog, fox or badger. The area this occurred in was at Hammerstream Paddock, Headcorn, Kent. The paddock has a railway line along one side, and a river the other side, and with some overgrown parts." **(Source: *BCIB*).**

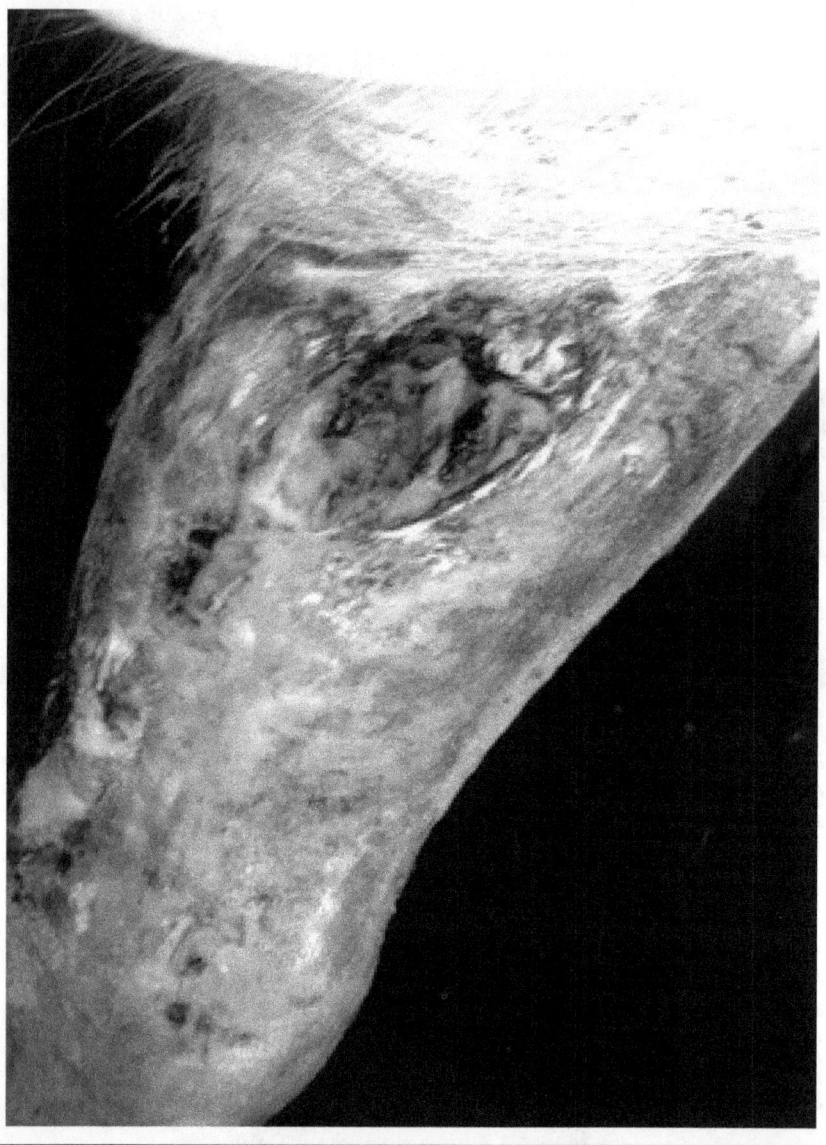

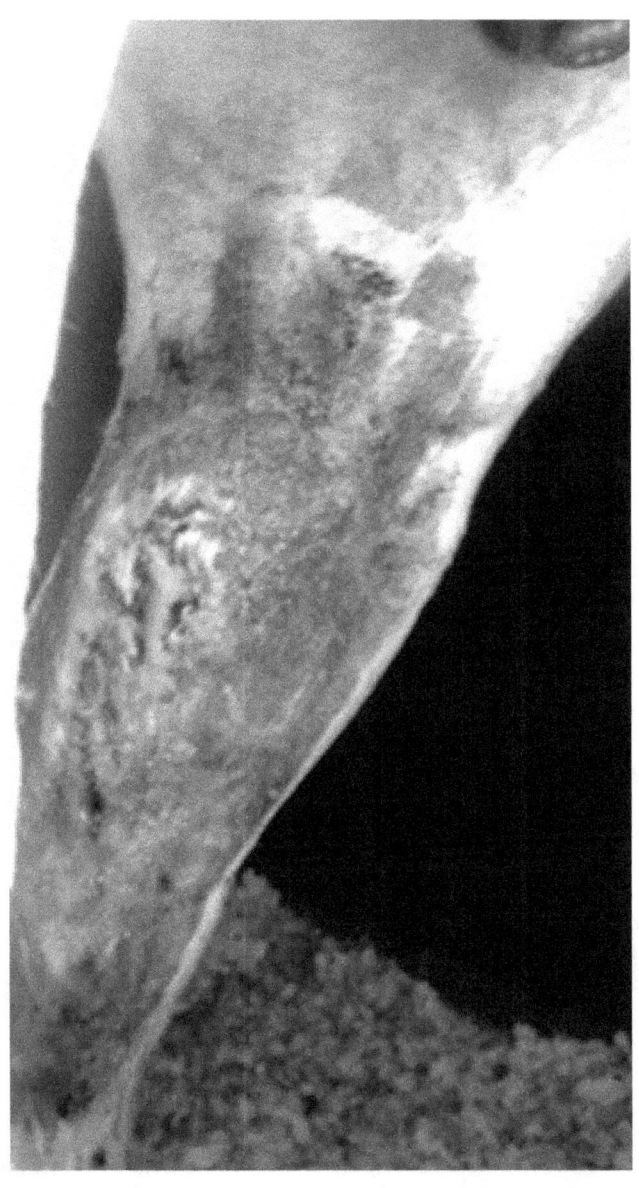

July approx: Train passenger Nicola Short says she saw a big, black animal descend from a tree and disappear into the undergrowth in Crayford. **(Source: *St Albans Observer*).**

Early October: Witness believes he saw a black leopard on the Isle of Sheppey, near Minster. **(Source: *Sheerness Times & Guardian*).**

14th October: A man claims he has seen a black cat "the size of an Alsatian dog" roaming the streets of Hildenborough. **(Source: *Courier - Kent*).**

18th October: Sighting by Justin and Ellie Ogilvy at 9:20pm, on Riding Lane, Hildenborough. The animal was all black, with a long tail, short hair, and was roughly 2ft long, and 18 in high.

They said: *"We were driving home one night, and the cat was walking along the left hand side of the road towards us. When it saw us it hesitated, and then jumped through the hedge into a near-by field. Recently, our two cats have been very nervous when going out into the garden."* **(Source: *BCIB*).**

Lancashire
8 reports

1st February: A young woman says she got the fright of her life when she saw a huge, black cat close to her home in Furness for the second time in less than a year. **(Source: *The Westmoreland Gazette*).**

19th February: *"At 23.00 hours this evening I was closing my bathroom window and there, not 10yds away, was a lion walking down my back street - Lonsdale Road in Bolton.*

The witness reports the next day: *"I have been looking at cats all day and the thing I saw was like a cougar / puma. It was about 7ft in all, and it had the stealthy walk and long tail, and was a beige/gold colour"*. **(Source: *BCIB*).**

11th March: *"Hi all. Just arrived home. At 4.15 p.m. we spotted a big black cat at Newburgh. Scouted round for half an hour but only saw it twice. Talk about an emergency stop. Going to look at photos on pc and see if anything shows up on them. It was no moggie!"* **(Source: *Cheryl Hudson - BCIB*).**

20th May: A two witness sighting at 1130 am in a field off Ford Lane, Goosnargh, Preston. It was black, and its tail was about 2-3ft with curve and quite thick. It was seen for about 1.5 minutes, at a distance of about 100m, and 2-3ft high, and about 6-7ft in length, including the tail.

Graham Smyth said: "*We were checking on some cows, when we spotted the cat across the stream walking across the field. It walked very slowly with a very smooth action. It was unmistakably a large cat. I tried to get a picture, but panicked a bit and used the wrong setting. The cat disappeared into the hedgerow.*

As it was walking, I stupidly tried to get the cat's attention by whistling, but it took no notice. I am very sceptical about these things so couldn't quite believe what we saw. My wife was with me at the time." **(Source: *BCIB*).**

18th June: *"I am writing after the events that occurred on the 18th of June 2007. I was walking in, and around, an area called Kirkham near Blackpool. I was near the wooded area of the village, when I heard a loud growl. I suddenly found myself leaping towards a tree, too scared to move, I waited for maybe three minutes before a large black figure emerged from a grassy area.*

It was jet black, and about the size of a Great Dane. It was probably 3-4 ft high, and it didn't look like a dog. I was only about 15 m away, so I was sure it was a large cat - it saw me and ran away. I was shocked that these large cats were so close to civilisation - about a mile away to be precise." **(Source: *BCIB*).**

Mid-July: Prints found at Formby. **(Source: *Christopher Johnston*).**

September: *"I had a job interview not long ago, in a small place called Padiham (located in the middle of a load of fields, on the outskirts of Burnley) and as I was taking the bus to my interview. I could swear that I saw not one, but two, big black cats (panther comes to mind, but who sees panthers in England, right?) in the middle of a field to the side of the road (private land is my guess, as there were a good hundred sheep in the field).*

What's interesting about that, although I was the only one on the bus, with the exception of the bus driver, who also spotted them, is that as I said, the field was full of sheep, so I could accurately judge the size of the cats (which were about twice the size of the sheep to my memory, well about twice the size in length, 1½ in height).

So yeah, I wasn't able to snap up any pictures, but they were practically in clear view to anyone who drove past. So I was curious as to whether anyone else has made any reports from this area. Just so I can confirm that I, and the bus driver, still have our sanity." **(Source: *BCIB*).**

Date and location unknown: more prints found. **(Source: *Christopher Johnston*).**

Leicestershire
36 reports

3rd February: About 18.00hrs hours on single-track road between Sibbertoft and Theddingworth, animal smaller than Alsatian crossed road described as Grey straw-ish colour with black rings? **(Source: *Neil Hughes PWLO*).**

4th February: *"It had to happen! Just a year and four weeks since the puma was last at Knossington, on the Leics/Rutland border, when it killed the sheep next to dad's house, it's back. Sally Knight and her husband, dad's neighbours, heard it outside their house in the same field as last year at 02:30 and 05:30* *this morning. Sally turned on the lights and it stopped making a noise, and started when she turned them off. She described it as a cat-like scream like something from a horror movie. It was exactly the same noise they heard last year when the sheep was killed.*

Sally spoke to people during the field trip, at the conference, and was interviewed on TV. Also, both my father (yesterday) and his handy man (last weekend) have heard a cat-like meow noise, but very loud, travelling over the field behind. He thought it was like a kitten, but much too loud. They both searched in vain for the source. Dad has located a 10-foot high woodpile with deep scratch marks all the way up next to the barn in that field. Plan to keep close eye this weekend and see if it appears again." **(Source: *Nigel Spencer RLPW*).**

11th February: Nigel Spencer reported: *"I have just taken a call from Colin Mayes, one of our investigators in the West Leicestershire area.*

Today Sunday 11th Feb, at 10:30 this morning, Colin was walking up the track leading to the top of Bardon Hill (at 278m, the highest point in Leics and Rutland) which he does regularly, due to the high number of big cat reports in that area dating back many years. Next to the hill is Bardon Granite Quarry (Bardon Aggregate Industries) and one of the site workers driving up in a land rover stopped and spoke to him. Colin made no mention of what he was doing, but the guy asked if there was any unusual wildlife up there, and he went on to explain that he had just seen a very big black cat in the woods at the foot of the hill by the track, much bigger than a fox.

He was quite shocked when Colin explained the history of sightings on the hill!" **(Source: *Colin Mayes & Nigel Spencer RLPW*).**

25th February: Nigel Spencer of the *Rutland & Leicestershire Panther Watch* received the following email: *"I'm ----- ----- and I farm at -------, Alverton at the top end of the Vale of Belvoir, just North of Bottesford.*

Yesterday I was opening up a ditch at the back of an old manure heap and came across the carcass of a black cat. It had probably been there for a month or so but still had most of the coat intact.

Its tail was about 10 in long and the body a further 22 in, and when alive it was probably 15-18in high. I only bring this to your attention as we have had two big black cat sightings here previously. These occurred in consecutive summers in 2004 and 2005, probably July. In both cases, the cat was disturbed by my son taking our Airedale for a walk. In 2004, it was on the disused Newark to Bottesford railway, when our dog surprised the cat, which my son (then aged 22) said was around the size of a spaniel- about 3/4 the size of our Airedale. The second sighting (2005) was very close to our house. On returning from walking the Airedale, my son found the cat apparently sleeping under a hedge, which bounds our rear paddock. The cat promptly leapt over a 3ft 6in fence into our neighbour's very overgrown orchard. My son rushed into the house and we went straight round to the neighbour, and together we searched his orchard, but found nothing. However, the area of flattened grass on the south side of the hedge showed where it had been sleeping in the sun.

I have left the carcass in situ *for the moment should anyone wish to see it.*

Nigel contacted Rob Cave who hotfooted it around to the farm and retrieved the rotting carcass and brought it home. He snipped off a few pieces of fur and sent them in the post to Mark Fraser at BCIB.

Rob's initial report is as follows:

"It is as stated – approx. 22in long in the body and 15in high. It has black fur and is crawling with maggots. Tail not long and hooked. I am arranging for an expert to look at it and sending fur to Mark for DNA analysis. "The animal had a broken and worn front tooth and was found laid out in the bottom of a wet ditch. The maggots were white/cream, and approx. 15mm long - if we have any blowfly forensic experts here.

Temperature around these parts has been 4 degrees to approx 10.5 degrees lately. I'd guess at 2 weeks to a month dead. I grabbed a neighbours overfed kitty and measured it after I got back and it was also 22" long in the body, so it is possible it could be a domestic gone wild. Bought a disposable camera and gloves and bags and went to scene. More info will follow after follow up actions have been completed."

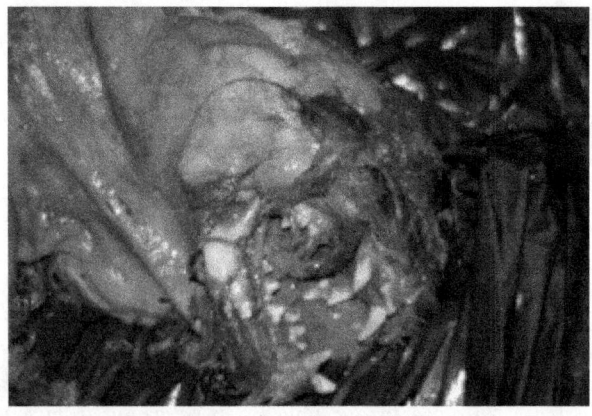

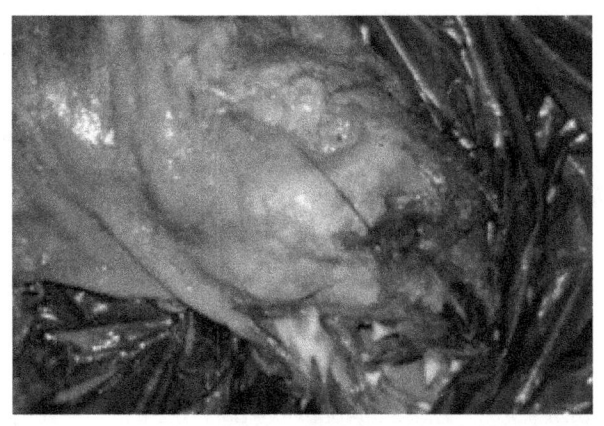
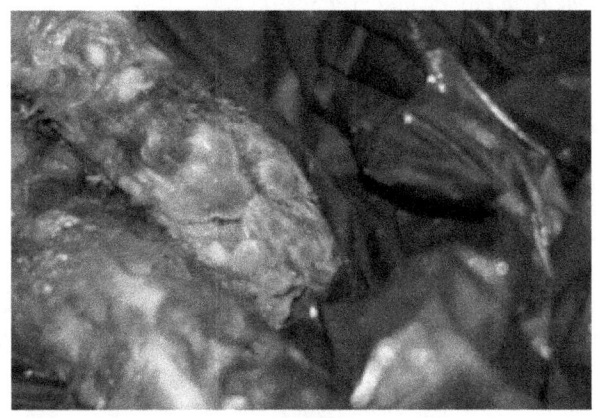
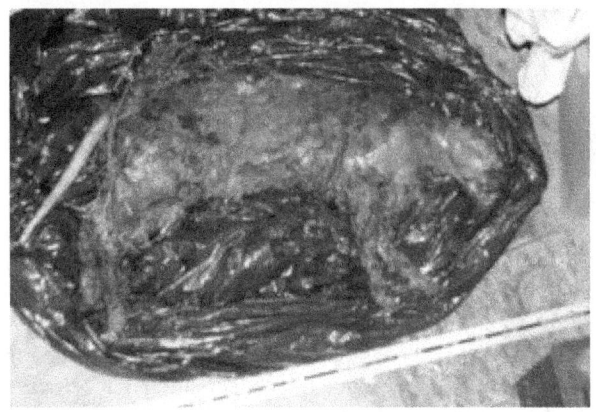

Rob adds: *"I've measured it from the back of its neck to the top of its tail. Nose tip to tail end I reckon is around 41in. The tail is 1 in, and weighs about 15lbs. Nail colour white: lower front teeth damaged. Fur essentially black but very decomposed.*

I contacted Twycross Zoo first thing this morning as they had previously indicated a willingness to look over anything we brought in. They promised to call us back: at noon I rang them again and got the same message.

I sought Nigel Spencer's advice and contacted Robin Roberts - Head Keeper at Drayton Manor Park Zoo near Tamworth. Fantastic chap. We took the body over, and not only were seen immediately, but had a private tour as well as a really good natter with a man who really knows his animals.

In Robin's own words: 'That's a bloody big domestic cat!' (Probably the biggest domestic gone feral that he had ever seen).

He did however recommend following up with DNA in case there is anything else in it. (DNA sample already on route to Mark) **(Source: *Nigel Spencer RLPW & Rob Cave*).**

4th March: Sighting at 9.30-9.45am in Coalville-Whitwick. The animal was black, and its tail was very long, not like a dog's. It was spotted for about 3 minutes, at a distance of about 100-200yds.

The witness said: *"Me and my Stepmother were on our way shopping when she looked over to her right to see two 'big cats' running across the field. They were running too fast to be large dogs, and their tails were curled at the bottom."* **(Source: *BCIB*).**

11th March 2007: *"Hi everyone... (Tonight-about 7-30) I spotted the cat whilst I was out walking my German shepherd. I was walking along the mud track looking out for foxes along the railway line (I was below it) when I spotted an animal sitting there watching me. As I got a bit closer, I realised it was a large cat about Labrador size. It was getting dark so the colour I am not sure about. But I know it had a big head - I noticed that straight away. When it got up, and I lost sight of the cat, I decided to leave the area -very quickly I fact - I wasn't brave enough to hang around. The sighting lasted for about 5 minutes -it was watching me and my dog, but every now and then it would look down the train track."* **(Source: *Donna Brown*).**

20th April: PC Kaiser was jogging along a former railway line in Shepshed at 12.50pm when, he says, he saw a black cat around 4ft long and 2ft high. **(Source: *The Leicestershire Mercury*).**

23rd April: At Shepshed, Loughborough, 10:30pm. Nothing was seen, but a very loud cat-like growl was heard. The witness said: *"When getting the dog in from the garden I heard a big cat-like sound, which you would hear on a nature program. I made a quick retreat to the house. I didn't think to report it until I saw the report in the local paper (Leicester Mercury)."* **(Source: *BCIB*).**

24th April: Police officers say they have spotted what they believed to be a big cat during a morning patrol. Traffic officers Pc Darren Elsom and Pc Louise Raven say they saw the animal loping across a hill as they were driving through Ellistown, near Coalville. **(Source: *The Leicestershire Mercury*).**

26th April: *"I have had my second sighting of a large black at about 7.30 last night. At first I thought it was a bin liner. I went home and got my dog, then returned to the area, and found flattened grass, and scratch marks up the tree. Close-by, just over 6ft up the tree, twigs on the tree had been broken, and very close to this is a small tunnel that leads under the railway line into some school fields which has had a sighting of a black cat."* **(Source: *Donna Brown*).**

8th May: Robert Keeling, of Parklands Drive, Loughborough woke up last Tuesday to find a panther-like animal in his back garden.
(Source: *Loughborough Echo*).

13th May: Susan Webb was walking a few yards away from a disused railway line, near Charnwood when she saw a large black cat "preening itself."
(Source: *Loughborough Echo*).

15th June: *"Hello, I am speaking on behalf of my daughter who runs a pub called the King's Arm's at Hathern, Leicestershire. She saw what was a very big black cat going into fields at the back of the pub. She said that it looked like something between a cat and a panther - that is how she described it to me. This was spotted in the afternoon and she is adamant that it was too big to be a normal domestic cat."* **(Source: *Nigel Spencer RLPW*).**

18th June: Nigel Spencer reports: *"We are now following up a report from Blaby Country Park, Countestorpe Road near the hospital lane.*

An off duty policeman heard a deep growling from a bush there on Monday night - his dog panicked, and it unnerved him that much that he didn't feel safe to investigate further. He, along with others we are in contact with, are now monitoring the location with cameras ready!" **(Source: *Nigel Spencer RLPW*).**

27th June: It is a well-known haven for wildlife - but did David Garratt spot a black panther running through Aylestone Meadows?

David, 19, is convinced that he and his fiancée, Lucy Acton, 22, saw a big cat in Aylestone Meadows Par, Leicestershire, as they took dog, Poppet, for a walk. The couple, of Cheriton Road, Eyres Monsell, also found the remains of a large bird's leg, which they scooped up as possible evidence of panther activity in the area.

BCIB member Donna Brown reports: *"Referring to the mystery bird leg that was found close to a sighting in Aylestone Meadows, I can confirm it belongs to a turkey - my other half used to keep turkeys, hens chickens and a peacock; he looked at the claw and said straight away that it belongs to a turkey.*

Nigel Spencer said that boys were actually throwing the turkey foot around at St Andrews football ground, and that it had been severed by a knife! " **(Source: *Leicester Mercury - Donna Brown & Nigel Spencer*).**

3rd July: *"I would like to report a sighting of a panther, made by my brother about a week ago, on the canal tow-path, near the gas works on the edge of Aylestone Meadows. It was around 3pm in the afternoon, and my brother was walking his dog, and got within 300 m before the panther ran off.*

I have also seen a panther some years back, on the cycle path at the rear of Gilmorten Estate on the old railway line. I got quite a close look as it fetched me off my mountain bike.

I have also seen what I believe to be a lynx type cat in the grounds of the tank repair depot in Old Dalby." **(Source: *BCIB*).**

5th July: *"Just spoke on air about big cats after Radio Leicester News presenter Gina Bolton, had a puma run across the A46 this morning near Wymeswold / Six Hills at about 05-00. She described it as brown with very big cat like tail and over 2 ft tall. I did the live on-air bit at 07:45 with her and Ben Jackson, who said it was amazing we were talking about this to him over 10 years back and its still happening!"*

More info from Gina Bolton: *"I'm not really sure how to describe where it was - I was on the A46 heading into Leicester and just after I saw the cat, I saw a sign to Wymeswold on the left.*

It was just before 5 this morning. I'll have another look tomorrow and see if there's a way I can describe the spot more accurately." **(Source: *Nigel Spencer RLPW*).**

12th July: *"A large unexplained animal jumped out of a tree and into a wood, making an unexplained noise, at approx 21.30hrs in front of my teenage son, who became very frightened. Can you contact us so he put some kind of explanation to this experience."*

Nigel Spencer contacted the witness and reports:

"I have spoken to this lady, and her late-teenager son and girlfriend, who were very upset by it, and he couldn't drive back home to Loughborough from Buckminste, which is out towards Grantham. He does a bit of lamping etc., and is used to things in the countryside, but has no idea what this thing was. It was bigger than a fox and had no tail, and was rabbit coloured. It moved like it was supernatural, and as it approached the wood, all the birds flew up. 'It was making a strange and scary cackling sound, sort of like a vixen'. He is going back with the Buckminster estate gamekeeper to investigate, as he is too scared to go back on his own. I have asked them to report it on the BCIB website." **(Source: *Nigel Spencer RLPW*)**.

2nd August: Sighting at Magna Industrial Park, Lutterworth, South Leics, when a couple were driving back to Market Harborough at 60mph. A big black cat-like creature, which was longer than the width of the car, and slightly higher than the bonnet, leapt across the road. They stopped, and it was about 2 ft. in front before it disappeared into the undergrowth. They described it as having a mottled appearance. **(Source: *Nigel Spencer RLPW*).**

2nd August: *"Whilst walking my dog on 2nd August at approx 8-45pm, in the location of Hugglescotes disused rail-line, just west of the bridge on Grange Road, I was startled by a loud deep growling noise with a distinct purr tone, behind me in the bushes. It's an area I use regularly early in the morning, and late at night - sometimes after dark - to walk my dog, and often encounter foxes etc. But never noises of this sort. Casting my mind back to 2005, I do recall seeing a pair of animal eyes reflected in my lamp one night, they were blue and quite far apart. Absolutely nothing like a fox or badger, and I remember wondering if it was possibly a big cat. I dismissed it as litter in the bushes. I'm now convinced I'm not going mad after all."* **(Source: *Nigel Spencer RLPW*).**

25th August: *"I was walking in the fields around the village of East Goscote where I live, and was walking through the field nearest to Broome Lane. This was 7pm yesterday, 25th August. In the field next door, I noticed something large and dark near the hedgerow. I stopped to look, and it was a large black cat-like creature with a long tail. It remained stationary whilst I watched it, but when I moved closer it jumped into the hedgerow. It was some distance away, but appeared to be much too big to be a domestic cat; it was also very muscular. I own a cat, and see lots of domestic cats when I am out walking, and this is the first time I have seen anything like this. I notice that there has been a recent sighting of a panther on the A46 at Six Hills, which is not far away from East Goscote. I intend to go back with my camera!"* **(Source: *Nigel Spencer RLPW*).**

31st August: A prowling big black cat similar to a panther has been spotted in Old Dalby. Daniel Davies saw the big cat cross the road near the school early on Friday, August 31. **(Source: *Melton Times*).**

4th September: Sighting 200yds outstide Fleckney. The animal was described as a black leopard and was seen loping across a field through a gateway on the Arnesby Road.

"The witness turned around and searched the area, but could find no trace of it. I believe he was on a pushbike so the exposure can't have been more than a few seconds. As he was indisposed, I gathered what I could from his wife. I understand from Nigel that they declined interviewing for the local radio.

Anyway, this area is being heavily patrolled for two rogue foxes who are playing hell with the pheasants and partridges, so if anything else is seen, it'll get back to me. I spoke with the witness' wife by phone - he is thrilled and now absolutely certain big cats are here.

The location given is slap bang where I equipped a gamekeeper with night vision equipment in May. Having cleared it with the witness, I have spoken with him and he will keep an eye out whilst he is out. He has been busy with the night vision - twenty-six foxes shot since the end of May. He will not shoot the cat (before anyone panics). Interestingly, one of his assistants was sent into an inaccessible coppice with pond and big tree after dark, and reported something strange with a 'bloody long tail up it', which quite spooked him. Will try to get to inspect the coppice at the weekend." **(Source: Rob Cave - BCIB).**

5th September: *"This evening, Wednesday 5th Sept 2007, I visited the Belper Arms pub at Newton Burgoland to play boules. On arrival, I found that the pub had been closed down by the brewery. I went to speak to some travelling people camped behind the, now deserted, pub. They said that the landlord had left suddenly, and had given them two young goats from the now defunct 'pets corner' to dispose of as they wished.*

We sat at one of the picnic tables having a coffee near to where the goats were tethered. At about 8pm the goats started behaving in an agitated manner. One of the ladies went to see what the problem was, then we heard her scream. Three of us ran to see what the matter was, and there was a black panther standing on the other side of the fence on its hind legs with the front legs hooked over a two and a half foot high wooden fence. It had its ears back and was snarling at the lady. Its tail was swishing back and forth like a cat. It was easily the size of a big German Shepherd dog and weighed about 70 - 80 pounds I would estimate, but it had a cat-like face, and was jet black. We could see its teeth bared. We all started shouting, four of us, and it dropped down on to all fours and slung off into the hedge nearby. It was obviously after the goat's kid, and didn't like it when we disturbed it. We all saw it again later, slinking across a nearby field towards a herd of cattle, but then it dropped into a ditch and vanished.

I know that it was NOT a domestic cat, but the strange thing was it didn't immediately run off when we shouted at it. It walked away slowly from us, almost as if it was used to people. At least one customer who turned up to find the pub empty also saw it. So there were five people who saw it close up. The others I was with are travellers, and have moved on, but they told me that they were moving to The Globe pub at Shackerstone, Leicestershire on the 6th of September because the brewery had told them to move because they were going to barricade the entrance to the pub car park to stop other travellers from going on to it. The travellers said they would verify the story if necessary." **(Source: *Nigel Spencer RLPW*).**

12th September: - 3 sightings Ongoing ***situation in South Leics around Fleckney - report by Nigel Spencer*** – *"Following a 'phone call regarding a broad daylight sighting of a large black Alsatian-size cat in a field next to the village, I got Rob to check with his gamekeeper friends over there, and it turned out that there were two more independent reports within yards of each other last week ½ mile away that we were unaware of. They had also seen unusual creatures (large animal up a tree and very long thick black tail going into a hedge).*

As luck would have it, the BBC were going over Sunday night to film a feature on lamping foxes and night sights with Rob and the gamekeepers on a large estate in that area so they were keen to combine the two subjects. Then, yesterday, the gamekeepers came across a middle-age couple near a large coppice on the land that had just seen a very large cat go into it. The tail was white tipped, and the animal was black.

At the point of entry, the gamekeepers said there was a very strong smell, but not anything they recognised, and their gundogs went crazy at that point.

With this in mind, we split into three groups, each in a landrover, to surround the wood last night after dark. However we got pinged by a laser night scope, or similar, ½ mile away and after Rob's mate night-scoped the edge, there was an armed poacher present. This ruled out the use of any fox lamping, and had disturbed any chance of a big cat sighting.

After scaring the life out of whoever it was, by approaching covertly and setting the dogs in, we spent the rest of the night, until the early hours traversing the fields around the estate to no avail.

However, it was all very positive and, hopefully, we now have another group keeping an eye open for us, (all ex-police officers as well)."

Follow up by Rob Cave:

"We got some feedback from one of the gamekeeper's friends. He has seen a largish black cat with a white tip to the tail. He has confirmed it to be a very large feral. When the bikers reported seeing the cat enter the wood next to which we filmed, one of them mentioned the cat's tail having a white tip." **(Source: *Nigel Spencer RLPW*).**

15th September: A gentleman telephoned BCIB to report strange paw prints in his garden at Enderby. Rob Cave visited and has taken pictures. They are 3in across in irrigated soil in a chap's garden near the M1 J21, and Leicestershire Police HQ. Rob thinks they are not dog and they have no claws, but sadly, they had degraded when he got there. BCIB have not yet seen the pictures taken. **(Source: *BCIB - Rob Cave*).**

16th September: *"We followed up on a call to the Spencer's that a black cat the size of a fox had crossed Gartree Road into Turnbury Road at around 21.45pm tonight. The location is 75yds from Evington Golf Course, which enters open countryside to the East of Leicester. Also 100yds to the west is a stream with heavily overgrown banks (large trees) running up the back of Allendale Road into the Stoneygate area. We didn't find anything, but equally saw no urban foxes, and no domestic cats out. Nearly all properties here enjoy hedged gardens - some extending well back from the houses. Not somewhere I can explore without upsetting people."* **(Source: *BCIB - Rob Cave*).**

1st October: Sighting by two witnesses of a black cat with pointed ears. Its tail was very long and thick, and curved towards the end. It was about 25 in high and 3 foot long, and described as a large unusual feral. The cat was there for a couple of minutes and was about 300yds away.

One of the witnesses explained: *"I'm working at the village hall and was just on a tea break. The cat looked low down and looked like it was prowling."* **(Source: *BCIB*).**

8th October: *"I have just been told that the parents of a local police desk officer saw a cat for the second time on a footpath at Aylestone Meadows - around two weeks ago. Apparently, the cat walked across the footpath some yards in front of them and their dogs. Their border collie 'Bedlington' looked at the cat, and it looked briefly at them and continued on its way. The only description is that it is bigger than the border collie."* **(Source: *Rob Cave*).**

9th October: *"At about 8.15am this morning – 9th Oct 2007 – something flashed across the road about 80yds in front of me running between two fields. I could only describe it as 'cat-like', dark and extremely fast! This took place on the gated road that links Garthorpe and Waltham on the Wolds, nearer the Waltham End, just before you reach the first cattle grid as you travel towards Waltham."* **(Source: *Nigel Spencer RLPW*).**

9th October: *"I just spoke with dad's neighbour (who was interviewed in 2006 during the conference tour of the area) and she says they heard the same animal again on Tuesday night in the field next to their house, and dad's at Knossington, on the Leicestershire/Rutland border.*
The cat made the same sound the night the sheep got killed in Jan 2006 (shown at the conference and on 'Heart of the Country'). It seems to match a puma,

which my wife saw crossing the road nearby a day or so later, back in Jan 2006. It did come back earlier this year when Rob and Kerry Cave did a night vigil the next night there, with myself, to no avail." **(Source: *Nigel Spencer RLPW*).**

4th November: *"On Sunday, 4th November at approx. 4.15pm we saw a large black cat near Stapleford, Melton Mowbray. My husband is a deerstalker, and quite used to assessing distance and size of animals, and is sure that this was not a domestic moggie, and most definitely a cat. It was black and alone, walking up a hedge side. We were on the road from Melton and it was in the field on the left of the road, in front of the spinney shortly before the self-storage place. I hope that's clear! I was driving at the time, and didn't see it, but Mark is convinced that it 'was what it was'. He's field craft experienced and sceptical of this sort of thing, so I'd be pretty sure of his ID!"* **(Source: *Nigel Spencer RLPW*).**

16th November: *Location withheld* – *"Had a third-hand report of a sighting of a large black cat stalking a lamper's Labrador near us, from a couple of nights ago. This sighting is within 2 miles of our motion camera so is not unexpected. The locations are very, very sensitive right now, partly because the sighting came from someone who usually carries a .243, and sheep kills in the area have also been mentioned."* **(Source: *Rob Cave BCIB*).**

18th November: *"Today a dog walker has just approached me to report a sighting she had of a large black cat a couple of days ago near Aylstone Meadows. She asked me to report it. She was taking her dog out for a walk along the footpath that runs along the side of the canal, and as she was coming round a bend on the footpath, she spotted the cat heading towards her. She said her dog didn't notice it, but she turned round and went back the other way before the dog spotted the cat. When she looked back, the cat was walking into the bushes on the embankment heading into the field. She described the cat has being black, but you could see a red tint in the fur when the sunlight hit the coat. It was the same size has her dog, which is a Labrador, and the same build, if not a little bit bigger. She said that, at first glance, she thought it was a black Labrador."* **(Source: *Donna Brown BCIB*).**

12th December: Sighting by Pat Dumayne at 5.30am., while driving to work. She was just approaching Magna Park roundabout on the A5, when a large dark object shot across the road in the shadows, moving very fast, and disappeared into the dark verge.

She said: *"I blinked in surprise, as it happened so fast, and it was gone. My colleague saw the object as well, and we both agreed that whatever it was, it was very dark, and a lot bigger than a domestic cat. The speed was very, very fast. It wanted to get across the road extremely quickly, whatever it was."* **(Source: *BCIB*).**

27th December: *"The 'Stamford Mercury' put me in touch with a golfer who* has found some prints in a sand bunker around the 13th hole. I visited the fellow and saw a mobile phone picture of the print, which was approx. 4" round. The heel pad isn't clear and the toes show claws. However, as it is on the 'up' slope that's ok.

The location is of interest, the bunker being due east of a grassed corner of the golf course beyond which is a wood. The fellow informs me that deer occasionally frequent this corner, and the paw prints occur simultaneously with the presence of the deer. Also, the ground in the corner was chewed up as though running deer had changed direction rapidly. The low sun would make this bunker an excellent ambush point into the corner at this time of year.

The informant is seeking permission for me to access the course with night-vision. Golf courses are very open, and particularly suit night-vision, and my theory is to find the deer and follow them at a distance. Also of interest - this golf course is north of Rutland Water, not too far from Exton, which I think houses an Amur leopard at the Rutland Falconry Centre, and also Barnsdale, which has seen deer kills in the past. I have put Nigel Spencer in touch with the newspaper and will contact the reporter next week to confirm the print is likely from a cat." **(Source: *Rob Cave*).**

Lincolnshire
7 reports

March 2007: *"We live in the village of Anderby (not Anderby Creek), which is on the east coast of Lincolnshire, 12 miles north of Skegness, 7 miles south of Mablethorpe, and 5 miles east of Alford. My wife and I are interested to know if there has been any recent sightings in this area.*

Three incidents have prompted this question. Several months ago I noticed very large pug marks in a field some distance from the nearest road, at the rear of our house. There was a distinct trail out from Seaton Farm, which is uninhabited, with many barns etc., and another returning in that direction. They were larger than a man's clenched fist.

On Wednesday, at about 3pm, my wife, who was at the sink in our kitchen overlooking the fields, saw a large black animal walk across the open pasture field immediately at the rear of our house, and only about 100yds from where I saw the pug marks.

Yesterday, I went for a walk over that field, and off to my left, just rising out of a ditch, was a large black cat like animal. It was about 300yds away from me. At first it was crouching down the slope of the ditch, and all I saw was its head - very cat-like. It came further into my view, close to the post of a stock fence, which is about 3 ft tall, and the creature was almost the same height as it. It was watching me very closely. Thankfully, common sense overtook me and I quickly returned to our house.

We are both down to earth, sensible people and not given to flights of fancy. We always believe that there is a logical explanation to most things in life. However, on this occasion we are lost for words. We are obviously concerned that if this is a wild cat-like creature it is very close to habitation, with the obvious problems that could create.

The fear of ridicule and being labelled eccentric cranks deters us from mentioning this to anyone else." **(Source: *BCIB*)**.

16th May: Sighting by James Weller at 8:25am, at Hob-Hole Bank Drain nr the Pilgrim Fathers Memorial, Fishtoft, nr Boston Lincolnshire. The cat was a predominantly sandy colour throughout, with a distinctive white tip to a short tail. The tail was quite thick and stunted, at around 20 cm in length with the white tip to finish. It was seen for about 10-15 seconds at a distance of 100yds or so – it was hard to say.

He said: *"My memory, I think, is inclined to play a fisherman's trick on me. However, I was out walking my dog - an English springer spaniel, and the cat was around the same size, possibly larger, but significantly stockier.*

"I was walking the dog, as mentioned, across a field towards the Drain (Hob Hole Bank). On approaching a fenced area, which leads to a path adjacent to the drain, I saw, what at first, I thought was a muntjac deer given it's broadness and white tail, which you commonly see on this walk. The animal was clearly aware of my presence and calmly skulked into the hedgerow and scrub, and I lost sight from there. In that time, however, my first impressions of it being a deer where quickly changed when I noticed the tail was longer than that of a muntjac, and wider and furrier. And its gait was akin to that of a larger cat. It was well muscled throughout, with what looked to be powerful limbs much larger than that of a wild cat and bigger in size." **(Source: *BCIB*)**.

6th June: Sighting by two witnesses at around 16.50hrs, near Holton Cum Beckering / Lincoln Road. They did not see a cat, but found a huge stool, plus dead rabbits (three in total - all with there insides removed) plus a large footprint.

One of the witnesses said: *"I was shooting rabbits as part of pest control asked for by the owner of the land. This cat has been seen many times, as it is on a campsite that is very quiet. I do all the pest control on the site and am there every day... I shall try to get some pictures."* **(Source: *BCIB*)**.

6th July: Prints found in an undisclosed location. **(Source: *BCIB*)**.

7th September: *"We thought that we should let you know that we saw what we thought was a Eurasian lynx tonight (7th September 2007) on Laughton Lane, between Gainsborough and Laughton, near Scunthorpe, Lincolnshire. We had a very good sighting of the animal, as it was on the grass verge to the left of the road, in front of some thick hedges. We estimate the size of the cat to be about 80-90 cms in length and about 60-65 cms high at the shoulder, it appeared to have quite long, slightly shaggy 'rabbit' coloured fur, with some dark markings, and pointed ears and quite a stocky, solid build. Weather conditions were good at the time and we saw the animal clearly in our headlights."* **(Source: Mike Bowdidge and Marcus Hammond).**

28th October: Sighting by four witnesses at Bayford Green. It was jet black, with quite a long tail and stood about a meter from the ground.

Sarah Siddique said: *"We saw it, and it was about 30 m away, and then it keep on disappearing and we kept on seeing it for about 20 minutes, then it would disappear. My sister and I were doing homework, when she pointed out a cat in the field. 'That's unusual', she said.*

I jumped to my feet realising that it was about four times the size of any small cat I have ever seen. It stood about a meter in the air and was jet black. Alarmed, I called my mum and dad, who raced to the scene. We got recordings and pictures of it.

I jumped on to the fence and saw it - 1 meter tall, and jet black, but it did not seem to see or realise I was there, even though it was about 30-35 m away. I did not see its eyes, or get a very good description of it claws, or facial features.

No other animals were there, but far way there looked like there was a carcass (of a sheep) but no others animals were near it. I made sure all my cats were in. As for the evidence, I don't know as I have not seen it, or been in the field yet, but it was today so I'll search for some evidence." **(Source: BCIB).**

10th November: Large black cat spotted near Horncastle. It left the witnesses with an eerie feeling. **(Source: *Horncastle News*).**

Middlesex
4 reports

4th June: Sighting by two witnesses at 19:20hrs, on Ferry Lane, Shepperton. The cat was black, but was too far in distance (approx. 70 ft) to see the height. It was observed for 2 to 3 seconds, and was possibly 5-6 ft in length.

Dean Prangnell said: *"We were driving down Ferry Lane towards the river,* when a black object, about 70-80 ft in front of us, raced across the road from one side to the other. As a cat owner of many years, I knew that this object was cat-like, also it was too big and fast to be a dog. The object had disappeared by the time we reached where it had crossed the road, and we couldn't find it, as the area is very dense in hedgerow and trees. The area is also mostly unpopulated with many fields and inaccessible/private riverside areas. Visibility was perfect as it was a warm sunny evening. Both myself and my 10-year-old son saw it, and were under no illusion that we had witnessed a 'big cat'. However my daughter was playing with her Nintendo at the time (typical)."* **(Source: BCIB).**

July: A large golden cat was spotted in a field near Radlett **(Source: *Enfield Independent*).**

Approx 1st August: Monique Jowers, of Kenilworth Drive, Croxley Green, says she was reading the paper at around 10am when a leopard-like cat attempted to enter her house through the French windows. She said it had orange fur with red spots. **(Source: *Enfield Independent*).**

7th October: A mysterious big cat has been spotted in Kings Langley. The large dark cat, with a big head and bushy tail, was sighted on Watford Road on Sunday, by an Enfield businessman, and a local farmer. **(Source: *Harrow Times*).**

Norfolk
23 reports

January: A large black cat was spotted running across the road, in front of a motorist, in the village of Hevingham. **(Source: *BCIB*).**

January: A sighting of a large black cat near a fishing lake on the outskirts of Hevingham. **(Source: *BCIB*).**

15th January: Sighting at about 10:40pm, on the B1111 between East Harling and Garboldisham. It was a light sandy/tan colour, and the tail is the thing the witness noticed the most, being very long at about 3 ft, and the same thickness all along.

The witness said: *"I was driving along this road - it was a very clear night and my lights are also very bright and were on high beam. This cat-like animal jumped on to the road from the hedge/verge area and stood for 2 to 3 seconds before leaping very fast, back into the hedge/verge. The tail of this animal was very long. There are usually small deer on this road, and I've always been careful while driving here, but I know this wasn't a deer or fox."* **(Source: *BCIB*).**

25th February: *"I have just got back from walking my dogs near Roudham Heath, which is a couple of miles up from Bridgham where I live. My two fell terriers need a lot of exercise, and the family and me are regulars to the area. About half three in the afternoon on Sunday 25th February I was heading down a track to an intersection surrounded by planted woods - the whole area is fairly uninhabited - army land and nature reserve to the north, Forestry Commission and Sheik Hemin (sp) estate (game managed) to the south.*

Something caught my peripheral vision to the left and I turned to see, 25 m up the track, the hind quarters and long tail of a large black cat like-animal. There are lots of deer, foxes etc. around here and I can assure you this was definitely thin and rounded like a cat's tail, and the legs fairly thin up to a broad thigh. I kept my dogs close, and gave it some time to head in whatever direction it was going, before heading back to my car. On the way back my terrier bitch discovered a dead juvenile muntjac with no obvious wounds to its body. I would be happy to assist in any questions you have, I can only say I have exceptional eyesight, and have studied animal physiology at university. Whatever this was, it was a fair bit bigger than my dogs and too big for a domestic cat." **(Source: BCIB).**

20th March: Sighting by Bradley Elvin at 23.35hrs, near 'The Swan' public house (Taste of OZ Restaurant) Ringland, Norwich, Norfolk. The animal was dark brown/black and its tail was long and fluffy/bushy, curving down then flicking up at the end. It was seen for about 10-15 seconds, at a distance of 100yds, and was possibly up to 1m to the top of the head off the ground. Length is unknown.

Mr. Elvin said: *"I was driving home from work at approx 23.35 on Sunday evening on 20th March. I was driving past the golf club, and I turned a corner and the cat was in the middle of the road just before the bridge over the river. It stopped and looked at me - it had bright white eye shine - then it turned and hopped in to the hedge on the driver's side of the road on the golf course. I drove up to the point where it went in to the hedge, but couldn't see anything so I drove off. I was driving home from work at the Mecca Bingo on Aylsham Road, Norwich."* **(Source: *BCIB*).**

March: Lalla Blackden spotted a large black cat while gardening at the rear of her home in South Bretton. The animal was 7yds away from the witness. **(Source: *EDP*).**

March: Sighting at 4.10am on Watton Road by UEA Black. It had a very long tail. The witness said: *"I thought it was a fox at first so I stopped, and saw it was a big black cat. It just looked at me in my car."* **(Source: *BCIB*).**

15th April: Three people spotted a 'puma-sized' black cat in the West Runton area of Norfolk. David McQuirk received a phone call at work from a worried resident who had spoken to one of the witnesses, but she didn't know much more. **(Source: David McQuirk)**.

15th April: A lay horse rider rang Mark Fraser to report a sheep's carcass found in Norfolk, but she would not give the exact location, or let us see the pictures (different witness from the 14th). **(Source: *BCIB*)**.

12th May: A large black cat spotted near Thelveton on Tuesday. **(Source: *Diss Express*)**.

14th May: A large black cat spotted near The Heywood at about 6.45am in fields as you come from Diss. **(Source: *Diss Express*)**.

24th May: Barry Dyer spotted what he said was a 'black panther' on the A1066 near Garboldisham. **(Source: *Diss Express*)**.

25th May: Richard Thorne contacted the *Diss Express* and reported: *"It was early in the morning, just before 6.45am. I was driving through the area on my way to work, heading towards The Heywood, when I saw something moving across one of the fields to my left. Initially I thought it was just someone's dog out for a walk with its owner, but there was something about the way in which it moved that was just not right for a dog. I am sure that it was a big cat and, as it was black, I can only assume it was a panther."* **(Source: *Diss Express*)**.

May: Large mysterious cat spotted near Thelnetham. **(Source: *The Diss Express*)**.

May: Large mysterious cat spotted near Burston. **(Source: *The Diss Express*)**

May 2007: Prints found near Riddlesworth thought to belong to a big cat by the witness - dog like. **(Source: *The Diss Express*)**.

3rd August: Sighting at 21.15 hrs by Steven Galley and another witness between Thursford and Little Snoring on the A148. They saw a dark brown, long slim animal, possibly 2.5 to 3 ft long. They described it as a large unusual feral, and the cat appeared to be the size of a fully-grown fox. They saw it for 5 - 8 seconds from a closing distance of 200 ft or less.

Steven Galley said: *"While driving towards Fakenham on the A148, a large cat crossed the empty road ahead of me and casually turned and walked back to the verge. As I approached the spot I quickly looked to the side, but could not see the cat. Although we were some distance from this cat at dusk, visibility was perfect and we both agreed it was a large wild cat that we saw."* **(Source: *BCIB*)**.

August: *"My colleague, who does quite a bit of rough shooting & coarse fish-*

ing in South Norfolk, was going home from our place of work at Occold, near Eye at, or about, 16.45 hours one afternoon in August.

As he passed the meadow by the River Dove, on the right hand side of the road, just before the road bridge at the end of Lowgate Street, Eye, he observed a big black cat in the meadow. He says it was the size of a Labrador dog, but longer, with a long tail and small head. Possibly this is the one that appeared at Cratfield." **(Source: *Rick Minter*).**

1st September: Sighting at 18:05hrs at Toft Monks/ Maypole Green of a golden yellow coloured animal, with clearly visible dark spots. The witness didn't notice its tail as it was lying down, but the body length was approx. 2 foot. The witness saw it for approx. 10 seconds, and was about 5 m away.

The witness said: *"I was cycling home from a friends and I noticed a large cat very unlike a domestic cat. It was clearly visible at the edge of an open field. It was eating a carcass that looked to be that of a hare or large rabbit. It had its head bent down into the prey and as I cycled by it raised its head for a split second, and then resumed eating. I did not feel confident enough to stop and have a closer look at it."* **(Source: *BCIB*).**

17th November: Sighting at approx. 4pm in Norwich, Norfolk, behind The Beeches Hotel, near the Catholic Cathedral. It was seen just above the old quarry which is now a sunken garden behind this hotel, and was black, with thickset limbs. Only part of it was visible, but it was certainly larger than any domestic cat, and appeared to be about 4 ft long. It was lying down, and the head or tail could not be seen clearly - it was also getting dark. It was seen for about 3 or 4 minutes and was probably 30 or 40 ft away, on top of a flat shed or garage roof.

The witness said: *"I was in my hotel room. I noticed this animal on a roof below my window, but across a gulley, which is part of a sunken garden. It didn't move, but as I turned to switch off the TV, and pick up my camera it disappeared. I am as sure as I can be that it was a very big cat, thick set and thick glossy black coat...unfortunately only part of it was visible. This was in almost central Norwich, so it seems unlikely...however I thought it was worth mentioning to you."* **(Source: *BCIB*).**

10th December: *"I was driving to Norwich this evening with my mother around 7:00pm on 09.12.07. We were just entering the fringes of Foulsham, Norfolk when a very unusual cat crossed our headlights. Its feet where larger than an average cat, its ears pointed and pink in the centre. The colourings a sandy yellow with large round spots - I believe it may have been a young bobcat or a fully grown spotted cat."* **(Source: *BCIB*).**

23rd December: Seen by Mark Crouch at 1.30 pm. It was jet-black, and had rounded ears. The tail was not as long as its body, but was a very thick tail. The witness aid: *"The thickness was the main thing I noticed, long and thick, with a little curl at the end. The height is hard to know as I could hardly see its legs, even though it was in clear view, but even without seeing the legs the height was 2 ½ ft tall – the same height as my German Shepherd dog. Its length was long, longer then the German dog - more like 5 ft I am thinking, as slightly longer then my other dog, an Alaskan malamute. The other thing was the oval shape - very oval, and it looked well.*

Two minutes is the sort of time that is enough to see this was a big cat and not a house cat or dog. I saw it from a distance of 20ft. I was walking my two dogs, Rocky and Max, at 12.00ish on the common at West Winch not too far from the train track. It was very foggy and was ice cold. I got to the end of the field and turned around to go back to the car - this was a 45min walk to that point. On the way back I heard a loud, very horrid scream, like a movie scream. So I popped my ears out of my woolly hat to hear if it was other people walking dogs. Mine were off their leads so if they get closer, I could call the dogs back, but I came across no-one and thought it must have been my ears.

Well, as I'm walking, and dogs are ahead of me, I came to the next field. Then from the bush this (dog) jumped out and ran on the field in a zig zag. I called the dogs straightaway thinking this was the people's dog I heard. I looked up to see it, and it was as still as day, just like a statue in the middle of the field. My dogs did not have a clue, it was looking at me black as hell, and I could not pick up what the face looked like as it was all back and foggy.

Well, it was there a while so I scaled it up with things around it, and this was no dog or house cat. I got the collar of one of my dogs, and by that time it made its way to the hedge still looking at me, so I made a big sound as I was holding my dog, and a few leaps that thing was gone. It headed towards the back of the homes and farms along the common. I wasn't happy, and unsure what was going on, so I got home to tell the other half and kids.

They did believe me as, after telling them, I looked this sort of thing up and the noise was that of a black panther - well that's all. I think, as there is building work going on close to the train track, it went around it and came across me as they use the tracks to travel. Also, I think we need to be aware of them as a fact not fiction. The cat seemed like a cougar, but black as they have thick tails and short legs with small heads, but are rare to be black." **(Source: BCIB).**

Northamptonshire
6 reports

19th February: Sighting at 6.45pm, between Stony Stratford, Milton Keynes and Towcester, Northants. It was black with rounded ears, and had a long tail with square cut off end. It was seen for a few seconds at a distance of approx 80 m, and was the size of a retriever or German Shepherd, but definitely was not a dog!

The witness said: *"As my husband parked the car at the kerbside a cat shot across the road at such a speed in front of us I immediately thought 'something is chasing it'. My husband got out of the car and walked around the back of the vehicle. Having unbuckled my seat belt, I was in the process of getting out of the car when there it was in front of me. At exactly the same time I stopped dead to stare at it, the animal stopped dead in mid-stride and stared back at me.*

It was completely black, round head with small round ears, long lithe body, with big paws and had a long tail with a square cut off end. I realised I was looking at a very large black cat, and felt fear grip my stomach. In the few seconds that actually passed of us staring at each other, my husband arrived at my side wondering why I had frozen half exiting the vehicle. I half turned to him and said, 'Look at....', but as soon as I looked back it was gone. He said he didn't see it." **(Source: *BCIB*).**

February: Sighting at 11.55am, on the edge of the 'Pheasantry' as it is known locally. The animal was jet black, with a long, sloping down, tail, which then curled up towards the end. There were two sightings in total, of two minutes, at a distance of 500m decreasing on the second sighting to 400m. The size was estimated at large dog, but much longer including tail, and sleek long body.

"We were walking back towards the car when our attention was caught by the animal crossing a field, towards a copse of trees and undergrowth. It was moving at a brisk pace with some purpose and was still clearly visible when it reached the cover of the trees. It then disappeared from view, but reappeared when it broke from cover in the same direction it had come from, when startled by a group of walkers with a dog on the other side of the copse who were unaware of its presence. The animal had a bounding gait, made even more unusual by its long body and curling tail. We were unable to investigate its tracks because it was on private land. We are 100% sure that this was not a dog or domestic cat, because of its size from the distance viewed." **(Source: *BCIB*).**

12th March: Seen by two witnesses at 4.10pm. The nearest town is Daventry, 2.5 miles away. It was sighted in a field behind Badby Wood. The weather conditions were good with sunshine. The animal appeared to be dark brown, but the witnesses were unable to describe markings because of distance away from them. It was seen for 20 seconds at approx 100yards away, and had pointed ears, with a long tail. It was difficult to judge its size, but was bigger than a fox. One of the witnesses said: *"We were sitting outside in our garden overlooking Badby Wood. We had been gardening, but had stopped for a cup of tea. The cat was walking in a stealthy manner, across the field."* **(Source: *BCIB*).**

14th March: Sighting at 03:20 hrs, on the A47 between Duddington and Wansford on a dry and clear evening. There was no mist or fog. The cat was completely black with no markings, and its tail was 2-3 ft in length. It was seen for 5-10 seconds, first at a distance of 150yds, which closed, to 10yds. Its height was 2 ½ ft and the length of its body was approximately 4 ft.

The witness said: *"I was driving a large goods vehicle along the A47 eastbound at 03:20 hours on Wednesday 14th March 2007 - a section of road I cover at the same time four mornings per week. As it was dark, I was driving on head and spotlights. In the grass of the nearside verge at approximately 150yds away, I noticed the rear of a black animal whose head appeared to be in the long grass that was further away from the road.*

As I got closer I slowed down, as to be honest from what I could see of the rear and the tail I thought it may be a stray Labrador. As I slowed to around 15mph the animal lifted its head out of the long grass and looked at my vehicle. It was then that I noticed that it had luminous lime green eyes and was of feline appearance. It the turned to face the field and, what astonished me, most casually walked into the undergrowth and out of sight." **(Source: *BCIB*).**

14th May: Sighting by Mrs Amanda Smalley –at 9.30pm, pm the Oundle to Warmington/Peterborough road. The cat was black with pointed ears, and its tail was about two ft long, held down, but the tip was held up. She is unsure of the height but estimates the length at 4ft+.

Mrs. Smalley said: *"I was driving from Peterborough to Oundle on my way home to Kettering. It was dusk and was fine weather, and I was beginning to slow down as there is a small stretch where houses are, and a 40 mile hour limit. As I did so, something began to run out into the road and I slowed even more. The first thing I noticed was the colour, thinking what animal is so black, then I noticed the tail very long and curled up at the end. It then dawned on me that what I was seeing was a big cat.*

Following the line of its body, I could see, as it was running, its muscles at the top of its legs - they looked so powerful yet it seemed to glide across the road. Its body was long and its coat seemed to shine in the headlights. On the other side of the road it ran into the ditch and then, I think, into the fields." **(Source: *BCIB*).**

25th December: Howard Moody was cycling to work in the evening when he spotted a large cat, known locally as the 'Beast of Bretton.' He said: *"I was by Lea Gardens, on my way to work at about 7.30pm, when a big cat crossed the path in front of me. It was definitely not a dog – you could tell by the way it was running – and it was far too big to be a normal house cat. It was three foot high and five or six foot long."* **(Source: *Peterborough Today*).**

Northumberland
8 reports

7th March: Sighting by Mrs Ann Brewis at 9.30am. She said: *"The sighting location was in our garden. It was black with pointed ears, but I am afraid the tail was behind bush, but it looked long. It was seen for 10 seconds from four ft away. Its height was approximately 3ft, and its body length was approximately 4ft, with the tail being extra.*

My dog (a springer spaniel) and I had just returned from our morning walk, and I was busy in garage. We are in a semi rural area, with one acre of garden, backing on to a field with overgrown woodland bordering our fence. The dog rushed up to the fence barking violently at a holly bush. I went over to calm her down saying nothing was there. I moved to the right hand side, and was confronted with this large black cat staring at me with yellow eyes just behind the bush/fence. It made no sound and after a few seconds, it started walking alongside the border fence into the undergrowth. I could here it breaking branches as it walked over the area. I then decided to get in quick with the dog as she was still barking violently, and I feared for our safety. I have lived in the house for 28 years. In fact there was one reported in the local paper last week, but this was 10 miles away, and I believe this to be a different one.

There were sightings last year across the road in a neighbour's drive at night, and there have been remains of deer found locally." **(Source: BCIB).**

14th March: Sighting by two witnesses from the King Edward VI School, Morpeth. They saw a fawn coloured cat with stripes coming from a black dorsal line, reaching about half way down the body. It also had black spots on the fawn colouring. The underside was white, and the ears were black and tufted. There was a lack of tail, but it was longer than normal. The sighting duration was three minutes, and was, at the closest, from about 5 to 6 ft. The animal was 18in at the shoulder, with a 2ft long body from nose tip to tail base – 4ft including the tail.

One of the witnesses said: *"We were staying at the school for the weekend as part of a re-enactment event. Between 5pm and 6pm we went outside to speak to somebody camped outside, and I noticed the cat walking next to one of our cars. I made a noise, and it saw me for the first time and visibly jumped. Then it slunk very quickly under the nearest car. I called my partner over and he saw it too - as a gamekeeper, and countryside ranger, he has experience of tracking animals, and also noting details of them quickly and accurately.*

The cat came out from the other end of the car, about six ft away from us, and paused briefly to look at us. It stood in sunlight, so we both got a very good look at its colouring. At this point, it bounded away at speed. We had been discussing what it was as the colouring was so unusual. We both also commented on its movements when running away, as it bounded like a large wild cat (the two halves of its body looked like they pivoted in the middle) with its tail straight up and kinked over slightly at the top. As it ran off, my partner pursued it across the school grounds, and at one point it turned and hissed at him, and then it dived under a hedge, which was the last he saw of it.

It didn't go near any other animals that we saw." (**Source: *BCIB*).**

10th April: Jamie Dixon and Jason Elliott were driving towards Halton Lea Gate when a "black panther" ran into the road in front of their car.

Jamie said: *"We were stunned,.but we didn't hit it. It was lightning quick. It was jet black and it was just like a panther."* **(Source: *Hexham Courant*).**

2nd June: Three witnesses saw a cat at 13.00 in woods to the east edge of Low Haber Caravan Site, Whitfield on a dry, sunny and warm day. It was completely black, with no other markings. The tail length was not unusual, in relation to body size. One of the witnesses said: *"It was difficult to see clearly, as the cat was walking away from the group on the woodland path as we joined the path. It was seen for half a minute at no more than 30m away. The top of the cat's back would be at, or just above knee level, I would say - therefore about 500mm. The cat was in proportion, not fat, and appeared to be in very good condition. It was certainly significantly bigger than any domestic cat I have ever seen.*

I was taking my 12 year old daughter and 7 year old nephew for a walk into the woods, which lie on a steep hill to the east side of the caravan site. A path leads from the site up through the woods, and joins a main logging trail (some 300m from the site).

We had just walked on to the path, where it enters the wood, when we saw the cat walking away from us up the path. The cat did not hear us at first and we watched as it walked several yards along the centre of the path. Within a few seconds the cat turned, saw us and then ran off into the wooded area. My daughter was so terrified she ran back to the caravan."

I have made this report as: -

a) No one took me seriously at the time.

b) This was my first visit to the area (visiting family). The family (who just bought the caravan a couple of weeks ago) mentioned what I had told them to another family and they were told that that they have made a similar sighting when out with their dogs.

Hope this helps. I am certain of what I saw and my daughter will con-

firm." **(Source: *BCIB*).**

2nd July: Sighting by two witnesses at 19.00 at Mountsett, near Hobson, Burnopfield. Lindi Hodgson said: *"It was solid black in colour with no visible markings, and had pointed ears. It was seen for about 10-15 seconds, from 10 m away, and was around the same size as my black male Labrador.*

We were walking our dog in the long grass near Mountsett crematorium. We were playing frisbee with our dog, when a large black cat moved through the grass (around 5m away from the dog). It had obviously been disturbed by our presence, and the cat was clearly visible to us as it stopped in our line of vision before running away by jumping through the long grass. It was somewhat larger than a domestic, or house, cat as it was a similar size to our dog (who is around 36kg)." **(Source: *BCIB*).**

July: *"I've had a very strange report, which has me completely baffled. There might be a cat involved. A friend of mine, who is a first rate naturalist, reported this encounter while he was fishing along the Tees at Girsby, near Sockburn, Darlington. From the other riverbank, he heard a gentle purring noise, not unlike what a moggie might make - though possibly a bit louder than typical - as he heard it across the river. Following shortly after that, and about 20yds down the river, he then heard a loud shrieking noise, similar to what a chimpanzee might make. The shrieking noise was repeated several times, over a period of a couple of minutes. However, there was no commotion as if one animal was killing another, he didn't even hear the animal, or both animals, if it was more than one, moving about. There was thick willow growth on the other side so he didn't see anything. Assuming that the shrieking was an animal and not the ghost of the legendary Sockburn Worm's last victim, my best guess was that the original purr was from a cat and that the shrieking was from something totally unrelated that was reacting to the cat. (My friend is primarily a birder so could rule out any bird species as doing the shrieking) Anybody able to offer a solution on the identity of the mystery animal(s)?"* **(Source: *Ian Bond*).**

17th July: *"I got a call tonight from a Police officer in Northumberland about a big cat sighting he had on Tuesday at 4.30am.*

He was driving slowly down a quiet country lane near Corbridge, when he saw a big cat sat by the side of the road with a rabbit in its mouth. The cat was as big as his Labrador and was brownish with stripes, like a brindled bull terrier. The tail wasn't particularly long though not stumpy like a lynx. He said it was just like a normal cat only six times the size; the rabbit was almost engulfed in its mouth. He stopped and watched it for 20 seconds from about 50ft. He was in a video car, but the video was switched off. He switched it on, but the sound of it starting up must have spooked the cat, which then leapt off and disappeared (if only, if only!). He got out to investigate and estimated the height of the cat from surrounding objects. He later compared this to his Labrador and came out at about the same size, i.e. about 26in.

I explained that no such cat was known to man (I forgot about clouded leopard). Anyway he was adamant about what he saw. There was the slightest possibility that what he saw was a puma with the stripes being the effects of shadows, though he thinks not, but everything else he would be prepared to give as evidence in court, as he put it. He is very interested in finding out more, so doesn't mind people knowing about this and would be pleased to discuss further, perhaps not the press though. He intends to do some more investigations, and a bit of field work himself.

I got the impression that this chap was pretty determined to find the underlying cause of this, so there may be more to follow." **(Source: *BCIB*).**

1st December: Paul Charlton spotted a mysterious black cat after pulling out of South Park off Eastgate, at noon. Several yards up the road, at Fellside, he spotted a large black creature about the size of a puma, jumping over a wall. He said: *"I saw the back end of a big, cat-like creature – its tail was curled and hooked – as it jumped over what is a 7ft high wall."* **(Source: *Hexam Courant*).**

9th December: *"I've just had a call from a lady who saw what she thought must have been a big cat in Northumberland last night, (9th). The woman was a passenger in a car with her husband driving along the A697, 400m south of the B3461 Alnwick to Rothbury Road at 10pm last night. She saw the animal briefly as it bounded across the road, and her husband also saw it as it was sat beside the road before it ran across. He thought it was a ginger colour, but she could only describe it as light coloured, greyish in the headlights, with no markings. It was a bit bigger than a fox. She estimated, by measuring something of a similar height in her house, that it was about 18in high, and certainly no less than 16-17in.*

It was muscular, and as it ran its tail went out backwards. The tail was fairly long, but was quite thin, rather than bushy like a fox. She described its coat as fur rather than hair to try and distinguish it from a fox (not sure exactly what she meant by this). Both of the observers were sure from the way it ran, and the texture of its coat, that it wasn't a fox." **(Source: *Ian Bond - Northumberland Big Cat Diaries*).**

Nottinghamshire
4 reports

26th February: Seen at 20.00hrs, in Toton. The cat was black with grey circles in colouring with pointed ears.

The witness said: *"I was on a late night bike ride with two of my friends when* we went round a corner and it was in the centre of the path. We skidded to a halt and it stared, slouched, and turned its head towards us hissing at us. It stayed like that until it leapt in to the high grass." **(Source: *BCIB*).**

14th March: A large black cat spotted running across the A1 near Coddington - Newark at 00.08hrs. **(Source: *BCIB*).**

17th April: Sighting by Richard Marson and Leiamara Marson at 1.55am, in Hucknall, Nottingham at the island at the bottom of Wighey Road, just after the Linby crossings. It was a dry, mild night and the cat was black and slender, with pointed ears. They never saw the tail and they saw the cat only for seconds from 6-8 ft away.

"We were driving on our way back home, and we were crawling up to the island looking at the rabbits on the left on the grass. One of the rabbits was about to cross the road, but all of a sudden it ran back into the bush, followed by a very smooth moving, long, black animal going at break-neck speed. It seemed like it was almost flying, as it moved so fast and smooth. Then it went into the trees, and the headlights caught the reflection of its eyes, and its pointed ears. Its eyes seemed to be green. The one thing that we did notice is it seemed to be very tall." **(Source: *BCIB*).**

11th November: *"Whilst visiting family in Worksop, I was driving along a road about a mile from the town centre, at night, which was still quite a built-up area, when suddenly around 10-15yds in front of my car, a fully grown black panther ran across the road, in full view of my headlights! It cleared the road in a couple of bounds! My 18-year-old son was with me at the time, and he said 'what the hell was that dad!' I knew full well what it was. It headed down a lane, which passes the old sewage works, and then along a canal. I would like to point out that I know my cats, and have had a very keen interest in the big wild cats of Britain for some years now. This is not the first big cat encounter for me. In mid-Wales, a few years ago, I came across a half grown puma, right on the edge of a certain small village, in broad daylight. It was stalking sheep, but was disturbed by me, and it ran off, went under a fence and headed towards woodland. Make no mistake about it, there are big cats on the prowl in this country, but, to be honest about it, what harm are they doing? There's no significant loss in livestock, or people. So, unless they become a real threat, let them get on with it I say! I would like to point out that I live in Anglesey, North Wales".* **(Source: *BCIB*).**

Oxfordshire
4 reports

28th February 2007: A big black cat seen standing on back legs playing with flapping plastic sheeting near Shilton, West Oxfordshire. **(Source: *Frank Turn-***

bridge).

4th May 2007: *"I have recently spotted a melanistic leopard in Carterton, Oxfordshire near Burford. I think it may have been the 'Beast of Burford'. Mr, Wey, who teaches at Carterton Community College, took a cast of a paw print. We think it was very young, and may have other family in the area. I saw it near the Carterton Leisure Centre. It was just sitting there, and it didn't notice me for reasons I do not know, but a passing car (not close to it) scared it away (this was at roughly 1.00 am)."* **(Source: *BCIB*).**

27th October: A mysterious big cat was spotted by a Thame man at around 10.30pm, in the countryside near Thame. **(Source: *Thame News*).**

14th November: *"I've heard the stories of big cats in Oxfordshire before and mentally dumped them in the same class as UFO sightings, ghosts and bogeymen; something for the weak-minded to believe in. Paranormal schmaranormal.*

Until last night.

I have no desire to be thought a crank, even though I saw what I saw, and heard what I heard. It's too easy to label people and, if you've not experienced something like this for yourself, it's easy to dismiss it. I cycle to work between my village in West Oxfordshire and Witney. I ride along the A4095 and have done most days for the last four years. I was riding home last night at around 6.30pm. It was dark. As usual, I crossed the little bridge that spans the old, disused railway line near Curbridge, and was heading towards Lew. As I drew parallel with a gap in the hedge on the right of the road, I heard a noise from the hedge that was, without any exaggeration, one of the most frightening things I've ever heard.

It was a low-pitched hissing, spitting noise, but very, very clear, and not the sound a domestic or feral cat would make. I also caught a glimpse of what appeared to be a large, dark, shape in the hedge. The noise was resonant enough to sound as though it had a very, very large animal behind it, and one that wasn't pleased to be interrupted doing whatever it was doing.

What was it? I honestly don't know. I certainly didn't see anything clearly, or hear any more than the hissing sound, but what I heard was absolutely terrifying in its intensity and its intent. Of course, it could have been a trick of the wind. I could have imagined the shape, although not the noise. Feral cat? Maybe, but if it was, it was seriously big. Badger? I don't think badgers hiss. Definitely not a dog. All I know is that I felt rather too much like prey for my sense of comfort. It's an experience I wouldn't want to repeat. I've taken pictures of some prints I found at the scene this morning - I've no idea if they're related to what I saw and heard, and they could be one of the local hounds, but you'll have a better idea than me I suspect!" **(Source: *BCIB*).**

Rutland
7 reports

21st January: *"I have just taken a report today of a very good big cat sighting near to Rutland water adjacent to Exton and Burley. It was at 15:15 on last Sunday, 21st Jan 2007, and was very close to the Rutland Falconry Centre at Burley Bushes. The lady, in Knossington, has lived, and worked all her life in the countryside. She said it walked across the road about 20yds away and crossed into the field, but had no urgency or concern and loped along in the field in flowing movements. She described the long S-shaped tail, which was blunt at the end. It had small ears set back on the head, with a very heavy-set face, with no prominent snout like a dog. It was jet black, and as big as a Labrador, but with a longer body. It had a very cat-like movement. She is 100% that it was not a dog, but a big cat. Interestingly, the Falconry Centre has a clouded leopard on a DWA permit privately. A panther has been seen in very close proximity many times over the last 10 years - the most recent being last January."* **(Source: *Nigel Spencer RLPW*).**

14th June: *"A friend of my son's father has seen a panther today near Wing in Rutland at 08:20hrs. He stopped the car, reversed back, and watched it for some time in a field as it stalked something. He said it was bigger than an Alsatian with a much longer body, cat-head and very long tail. Jet-black, he is adamant it was a panther."* **(Source: *Nigel Spencer RLPW*).**

17th June: *"I took a 'phone call of a sighting at Pilton/ North Luffenham. It was at 09:45 on Sunday morning when the cat crossed the road 50 ft in front of the witness on a narrow lane. This is under ½ mile from where the last sighting of similar-sized cat was seen on Thursday the 14th. The animal was black, and much bigger than a dog at about 5ft long, with long cat- like tail and head.*

I was there by 14:00 and found a local farmer in the next field repairing fences as the sheep had smashed through into the next cornfield, which he was surprised at. He hasn't lost any lambs that he knows of yet, although his gamekeeper says he will shoot it if he sees it.

The location is next to a deep cutting that used to be part of the old ironstone railway and quarry that takes in all the land in that area. Needless to say, the whole area is riddled with rabbit warrens." **(Source: *Nigel Spencer RLPW*).**

27th July: *"A chap, home on leave from the navy, rang to say that he spotted a big black cat this morning near Oakham, at Stretton, in Rutland. It was jet-black, and the cat was one-kilometre away. It had heavy shoulder blades, and was big and tall! It went into bushes, and the witness waited, but it never came out again."* **(Source: *BCIB*).**

8th & 15th August: *"I have just got back from Essex, and spoke to a postman who has seen a small puma-like cat in Whissendine, Rutland on Monday, and today, Wednesday 15th August. The first was by chance at 14:30, when he stopped in a lay-by on the Ashwell Road. He noticed a large brown cat sunning itself on the opposite farm track. He got out, walked towards it, and not until he was only 10 to 20 ft away did it get up. He then saw that it was about Labrador in size, with a very long tail. It walked off into some bushes.*
Today at 14:30 he stopped there again, and it was in the same place! He tried to follow it to get a photo, but it ran off at great speed. He says there is an old barn and farm machinery there so it may be using that as shelter. Having looked up on the internet, he said it matched a puma." **(Source: *Nigel Spencer RLPW*).**

18th October: Sighting by two witnesses at 00:40 hrs. The nearest village is Wing. The witness was driving along Morcott Road towards Morcott, and took the first left (single-track lane) towards Lyndon. The cat on the left, 150 – 200m down the lane. It had a long, black tail (approximately 2.5 - 3 ft) which was curved. They saw the back end of the cat going through the hedge and said it was about 30in tall. The sighting was over quite quickly, at probably no more than a few seconds.

One of the witnesses said: *"I rounded the corner into a small lane and saw the cat going through the hedgerow. It was only about 10m in front of my vehicle and was lit by the headlights. I reversed after realising what I had just seen, but it had gone through the hedge."* **(Source: *BCIB*).**

18th December: *"I think we have a problem with a big cat. We have not actually seen one, but have had two sheep killed overnight. Both of them had their stomach ripped out and their carcass was eaten down to ribs and backbone, so they only head and fleece left. In one, even the ears had been eaten off. The field is situated between Wardley Wood and Stockerston Wood, just off the B664 ½ mile from Uppingham, Rutland. We would appreciate any help you can give us."* **(Source: *Nigel Spencer RLPW*).**

Shropshire
15 reports

7th January: Students Alex Warren and Jack Mullock were travelling along a narrow road between Rushbury and Munslow when they saw a large, jet-black, cat-like animal near the hedge. **(Source: *Shropshire Star*).**

22nd April: Gary Sharkey, 41, and his daughters Ashleigh, 13, and 10-year-old Abigail, saw a large black cat from their car while on the A5223 Wellington Road on Sunday.**(Source: *Shropshire Star*).**

8th May: Cheryl Scarrott, from Hollinswood, reported seeing a big cat. She described the creature as far too big to be a domestic cat, after seeing it leap a 7ft wall. **(Source: *Shropshire Star*)**.

16th June: Sighting by Mathew Rowland at Shrewsbury Coton Hill at around 16.00hrs. It was [...] cold and wet. It looked like a lynx [....] and was more brown than black. It had a long tail, but the witness did not notice the ears. It was seen for approx 5 seconds in the opposite field.

Mr. Rowland said: *"I was with my dog. It was taller than my dog, and a lot longer. I was taking my dog for a walk, and I was walking down this hill, and I looked at the opposite field, and there was the big cat. My dog just stared. It didn't bark or anything, but just looked at it. I was very shocked and confused really, and after about 5 seconds it ran away. I think it saw me, but then it was gone. I was taking out my mobile, to take a picture, but it went.*

That was my first sighting I'm just shocked and don't really believe my eyes. I mean, what the hell is a giant cat doing in Shrewsbury?" **(Source: *BCIB*)**.

6th July: *"A witness reported his sighting via the telephone. He had just finished work, and was driving along the B5602 in Telford. He had just passed the 'Lambs' pub at Edgemond, and was heading towards the Harpur Adams College / Newport showground, when he 'experienced something that was really odd'.*

About 80yds from the college a 'thing jumped out of the grass verge and crossed the road in two bounds,' into a cut grass/cornfield on his left.

He shouted out to himself, 'what the ----' was that. It had a 'massive' tail - as long as the body itself - turned up and curled at the end. The cat, as he now believed that is what it was, had a 'big under-hang under its belly', coming from below the neck, and kind of 'fluffy skin.' It was as big as a Labrador but its head seemed small for its belly.

As the cat bounded across the road, it 'seemed to accelerate as it reached the grass hedge'. He stopped his car, and looked into the field, but by then the creature had gone. He didn't think to look for signs, and didn't want to hang around too long, as he believes he had just witnessed a black leopard running through Shropshire! Although he did return a couple of days later to try to locate any tracks, he assures us that on the day of his sighting, he had not just left the public house. He went on to say that two years ago, at the rear of his house runs an old railway line – the Shewsbury / Wellington line. He saw a large black cat trotting along the trail, 'ever so weird you know'.

He also mentioned that, in a local lorry yard (didn't know which one - recently apparently) the workers in the yard heard a noise, moved some sheeting, and a large black cat run out and away. This was described as having a long, deep body, but shorter than a dog." **(Source: *BCIB*).**

8th July: Sighting by Jamie Gould at the Craven Arms (not a pub) on the A49, at 6.30am. It was black, and had an S-shape tail, about 3 to 4ft long. It was seen for 40 - 60 seconds from about 10 - 15 ft away. It was 2 ½ft high, and 4ft long (not including the tail).

Mr. Gould said: *"I was out walking with a friend, and stopped for a call of nature, by a fence on the edge of a wooded area, when I heard a sound of breaking twigs on the ground. As I looked up, I could see the back end of a large cat, with a 3ft long S-shape tail.*

It went behind a fallen tree, and then I could see its large head looking at me through the branches of the fallen tree - it was in a crouching position. I then backed away a few paces and turned slowly around. The whole time I could still see the cat, staring at me. I walked slowly up the bank back to my friend, looking over my shoulder a couple times. The area seemed to get darker as I walked away, so I could no longer see the cat, I returned to my friend, and left.

I would like to know of any other sightings in my area. There was a sighting in January this year, at Rushbury, approximately 3 miles from my sighting. (Information on Rushbury found on internet)." **(Source: *BCIB*).**

1st August: Sighting at Churchbrime, Wynchurch at 19.00hrs, when the witness spotted a large black cat standing in the middle of the road. The size, when it was standing broadside, filled ½ of the road, (5-6ft long), and was as high as a Labrador. The cat walked off through the hedge. The witness reported the incident to the local police who took the sighting seriously. **(Source: *BCIB*).**

11th August: Sighting by Sheila Roberts at 8.20 am in Telford, on the B4373 between Heath Hill and Dawley Bank roundabouts, before the bridge, and heading towards Malinslee.

Ms. Roberts said: *"It was black with greyish patches. Though that may sound like a weird tabby cat, it was nothing like tabby. It was only as I drove closer that the grey was noticeable. It was the tail that made me realise that it couldn't be a domestic cat, because it was so long and thick. The animal was large - much larger than a domestic cat - which was what surprised me.*

It was also longer in the body than a domestic cat. I have never seen any of the species of cats you mention other than perhaps on TV, so relative size is merely a guess for answering this next one. It was a large, unusual feral seen for 10 - 15 seconds. I was driving on my own, towards home in Malinslee, and I had just come off the Heath Hill roundabout when I saw the cat emerging from the undergrowth/trees on the left, and walk straight across the road. It was walking with confidence and purpose, similar to how we see the foxes around here!

As I came closer (more slowly), it didn't even turn its head to watch me. He was on a mission! He then entered the undergrowth the opposite side. I didn't stop the car, as I needed to be home quickly. I have since driven past the area again today (Sunday) and notice that there is a small trodden pathway around where I saw him. Like the usual regular runways animals make. This cat, if it is like others seen, was minding its own business and we should let them be. However, it would be safer to know where they 'live', in order for child/ livestock safety. Perhaps they have always been around, but now communication is enlightening us all. A breed we are unaware of." **(Source: BCIB).**

22nd August: A man walking his dog along a busy Shrewsbury road spotted what he called a "black panther." **(Source: *Shropshire Star*).**

30th August: Train driver Ian Dowley spotted a big cat on his way back to his Wem home from Shrewsbury. **(Source: *Shropshire Star*).**

6th September: Sighting at 12:30pm in Ellesmere, near to Colemere. The cat was black in colour, and the tail seemed long.

The witness said: *"The cat that we saw was approximately the size of an adult sheep. We watched it for about 20 to 30 seconds, from approximately 150-200m away. We were sitting on a hill having a picnic looking out onto a wooded area next to a recent cut cornfield with a lake on our left. I remarked 'look at that cat', and my partner replied, 'that's a big bloody cat'. The cat was very low to the ground, and almost appeared to be stalking, and then disappeared into the woods."* **(Source: *BCIB*).**

September: A Bridgenorth man told the *Bridgenorth Journal* that he had recently spotted a large black cat, he said: "Me and my friend were rabbit shooting on land near to Henry Yates', and we both saw the big cat. It was in the same area that the other people saw it.
(Source: *Bridgenorth Journal*).

September: Roy Page, of Alveley, said: *"At about 4.30am on my way to work, just as I approached Jiggers Bank on the Ironbridge bypass, I saw a big black panther-like cat."* **(Source: *Bridgenorth Journal*).**

21st October: Sighting by four (three adults and one child) at Billingsley near Bridgnorth, the exact grid reference being so700865, whilst crossing a ploughed field. The cat was jet black, and was 2.5ft+ [...], and between 4 ½ to 5ft not including the tail. It was sighted at 500m then as they drove closer to 100m, they became more aware of what they were looking at. It was viewed for nearly 1 minute.

Mr. Chris Jones said: *"We were driving to Bridgnorth and saw a cat crossing a ploughed field. It stood out, as the soil was light brown and cat was black. My* wife thought it was a dog, then as we got closer we could see it was a cat. It seemed to be hunting two pheasants in the field border. The car in front stopped and took a photo on a digital camera which looked like a good photo. We stopped at the roadside and confirmed with photographer that it was a cat that he saw.

We were pretty scared and did not get out of car. It was hunting pheasant and they flew off as it approached."
Follow up by Martin Rainer

"*I visited the area this morning of the reported sighting near Billingsley, Shropshire. Unfortunately, nothing to report of any significance, despite looking round for about an hour. The area is mainly open farmland, with pockets of woodland. However, the further south I travelled on the road from Bridgnorth to Cleobury Mortimer there were extensive areas of extremely dense woodland. I contacted the witnesses yesterday, but have yet to get a reply. This road is fairly busy and it was virtually impossible to pull off, as there were no obvious spots to. I would like to know how two cars could come to a standstill (on a road that has heavy traffic despite its remoteness) to view the potential big cat. Hopefully, the witnesses can put me right on that. I was a little limited as I had my four kids in tow (who were all gutted that we didn't see anything!), but managed a quick walk in the field that the cat was possibly spotted in. No obvious signs of any tracks etc. But not surprising as the ground is quite firm. I hope to revisit at the weekend, hopefully after the witnesses have contacted me. This area is some 8 miles, as the crow flies, from Wharton Park Golf Club, which has had regular sightings."* (**Source:** *Martin Rainer BCIB*).

9th December: Telford Dairy Crest worker Shaun Yale was travelling to work at about 5.40am on Sunday when a large black animal crossed the road in front of him in Shawbury. Shaun said: *"The tail caught my eye, very cat-like, despite the size of the animal"*. (**Source:** *Shropshire Star*).

Somerset
37 reports

17th January: Stephen Rolph spotted a huge cat while on his way up Buncombe Hill on the Taunton side of the Quantocks. He said: *"At first I thought it was a dog, but as I got closer it was obvious it was no pet."* The 45-year-old told the Mercury that it was around 4ft long, with a 3ft tail and of a stocky build. It was no more than 30 m away. (**Source:** *Bridgewater Mercury*).

Early 2007: Maureen McCrory saw a big cat near Upper Weston, Bath. (**Source:** *Bath Chronicle*).

February: Sighting of a black cat (not sure, as it was in silhouette). The tail was

2ft long, and the animal stood at 2 ½ft at the shoulder and was 4ft long. It was seen for 45 seconds from about 20m away.

The witness said: *"I was walking my dog near Spaxton, and I saw a deer running at full speed, and then realised it was being chased by a big cat. I watched it until it disappeared then heard the cat growl."* **(Source: *BCIB*).**

24th February: Two friends, Paul Smith and Peter Coales, had been up in the Quantocks when they spotted what they say was a large black cat. **(Source: *BCIB*).**

24th February: A large black cat seen around the Royal United Hospital, Bath. **(Source: *The Bath Chronicle*).**

5th April: Irena Scavaracini reported seeing a large cat on the outskirts of Taunton. She said: *"It came out of the hedge and was bigger than a Rottweiler. I stopped the car and thought s***, what shall I do?"* **(Source: *Somerset County Gazette*).**

10th April: Dave Vowles has about 60 cows and a number of sheep, which graze on land at Upper Weston Farm. But in the past week, two of his newborn calves have been savaged, leaving Mr Vowles upset and mystified. Now he says he is looking for any information as to what has killed his animals - and he has not ruled out the possibility it could be a big cat. **(Source: *Bath Chronicle*).**

18th April: After a spate of black cat sightings around Weston and Corsham, a keen-eyed member of staff at the *Bath Chronicle* filmed video footage of a mysterious black animal. Vanessa Peters, who works for the newspaper's field sales team, was with her family at home in Sandpits Lane, Gastard, when her husband, Tim, spotted an unusual animal making its way across fields behind their home. **(Source: *Bath Chronicle*).**

April: Golfer spotted a lynx at Frome golf club - it emerged from a field at the bottom edge of Egford Hill. **(Source: *Somerset Standard*).**

25th April: *"I've had an email from a lady today that lives not to far from me (Simonsmith) about a big cat that visits her garden. She first saw it about a year ago, and she then went and planted a catnip plant in her garden. Since then every few days it comes back to her garden - it comes in, smells around and then just lies there."* **(Source: *Anthony Bevan*).**

April: Yeovil woman Wendy Pollard told the *Yeovil Express* that she had seen a large panther/puma cat sitting at the bottom of her garden in early April of this year. **(Source: *Yeovil Express*).**

2nd May: Sighting by three witnesses from Yeo Valley Foods, at Isleport Business Park, Highbridge. The cat was reddish fawn, and about 5ft long, and 2 to 3ft high. It had tufted ears, and a short tail, and was seen for 10 to 15 minutes at a distance of 100ft.

One witness said: *"I am a lorry driver and was on a pick up from Yeo Valley Foods. I was waiting in my cab on the loading bay. Opposite is scrub ground, fenced in tall grass bushes. I saw movement, reddish fawn colour, then I saw its head - it was a big cat, with short ears with dark tips, that were white inside. It had a dark tip to nose and was white under the jaw. It just ambled away from the fence behind some scrub. I got out the cab and got another driver, local to the area, and went to the fence, and spotted it again. We got another driver over who was local to the area, and we watched it as it made its way through scrub ground. It turned its head several times – cat head – and it moved like a cat hunting. Then it went over side of ground, and went along the railway bar-type fence. Its back was just visible above the grass and it was too big to go through fence, so it went under into some marsh ground with reeds. We saw it turn its head back, then it was gone. When up against the fence we were able to gauge the cat's length / height at 5ft long, and 2 – 3ft high – it didn't look to have much of a tail. One local driver said it was a cat too. I have not heard of any reports."* **(Source: *BCIB*)**.

4th May: Sighting in Shortwood Lane, Litton of a dark red to dark brown cat, with very slight barring, with pointed ears, and thin tail. It was about 1ft (not long), and was seen for about 45 seconds at a distance starting at 100yds, closing to 5yds. Its height was about 20in at the shoulder, with a small round head with pointed ears and a short nose, shortish tale, a long thin body, with long legs, which were longer at the back. Not a heavy animal.

The witness said: *"I was cycling up the lane (south eastward) relatively silently, at dusk, with my headlight on, and could see something in road 100 to 75yds ahead. As I closed on it, I realised that it was an animal in the middle of the road with its back to me. It was picking at a road kill carcass, and as I neared it, it turned its head and looked at me. Then, without rush, it got up and sloped off into a very thick hedge/thicket on my left. This was certainly not a fox, and was far too tall to be a domestic cat. It was about 200yds from the nearest dwelling. Apart from the thick hedge, there are open fields around the site with a wood three fields away.*

I walk the area very regularly and see lots of the wildlife. Two years ago, in an adjoining field, I (with a large torch) saw eyes at night, which were not explainable as deer, fox or badger. This sighting was as a result of my sheepdog making me aware of, and challenging, an animal I had not seen. My dog was very aggressive and barking madly." **(Source: *BCIB*)**.

13th- 14th May: A large black cat seen in Langport may have been responsible for the death of three lambs in Muchelney, according to the Western Gazette.

(**Source:** *Western Gazette*).

May:. A large black cat was spotted from 80yds away, which seemed to be stalking something in fields near Frome. (**Source:** *BCIB*).

1st June: Sighting by two witnesses, one of them being Adam Weightman, on the A37 approaching Midsomer Norton. It was midnight when the cat ran out from the roadside. It was very fast and agile, and jumped into a hedge on the opposite side of the road. It had dark reddish spots, a short tail - which looked very similar to a lynx – docked, but wide and floppy. The cat had pointed ears and was seen for ten seconds at a distance of 50m. It a similar size to a large dog. (**Source:** *BCIB*).

28th July: Sighting at 15.30 by two witnesses on a footpath behind 'Old Fosse Road' Odd Down, Bath. The cat was black with no markings, and had a long and thick tail. It was taller, and longer, than a domestic cat, but smaller than a panther. It was seen for 10 seconds fro about 30 – 40ft away.

One of the witnesses said: *"We were walking along a tree-lined footpath, towards Odd Down, with the Rugby Club field, and new housing development, to our left. The cat was moving towards us, then turned in profile, saw us and disappeared through the hedge into the Rugby Club field."* (**Source:** *BCIB*).

13th August: A teenager from Cheddon Fitzpaine claims he came face to face with a panther-like wild cat as he drove home in the early hours. He said: *"It looked at me, then just ran off into the hedge and disappeared."* (**Source:** *Somerset County Gazette*).

20th August: A builder at a Nether Stowey construction site (near Sedgemoor) spotted two big cats prowling the nearby fields. (**Source:** *Bridgewater Mercury*).

24th August: A lynx-like cat was spotted along Touches Lane, Chard at 4.00 am. The witness could not make out the colour, but thought it was a lynx as the tail was short, and it had tufted ears, although it was only a fleeting sight of 15-20 seconds. It was about the size of a medium dog.

The witness said: *"I was woken by a very strange sound, so I looked out of the window. We have an orange light flashing outside our house, and I caught a glimpse of something moving, and as I looked harder it trotted into the forest next to our house. There was a weird whooping type noise - we have also heard a screeching noise several times, but have seen nothing.*

We have had no other sightings really, just weird behaviour from our dog and the decline in deer sightings as we used to get them in our garden all the time.

BIG CAT YEAR BOOK 2008

If this is a lynx, we live near a wildlife park (Cricket St Thomas) and it could be that it may have escaped from there." **(Source: *BCIB*).**

10th September: Black, smaller than a cow, larger than a dog and very fast was how the latest sighting of the 'Beast of Banwell' was described by a resident earlier this week.
(Source: *The Weston Mercury*).

September: Panther sighting at Colfors near Frome.
(Source: *Marcus Mathews*).

September: Panther sighting at Colfors near Frome.
(Source: *Marcus Mathews*).

September: Panther sighting at Colfors near Frome.
(Source: *Marcus Mathews*).

15th October: Anne Mendelson caught on camera, what she believes could be a black panther. The animal was seen roaming in fields behind her home near Corton Denham. **(Source: *Western Gazette*).**

3rd (approx.) November: Helen Bishop saw a big cat twice while walking her dog in a field off the main road between Chard and Crewkerne.
(Source: *Yeovil Express*).

4th (approx.) November: Helen Bishop's second sighting
(Source: *Yeovil Express*).

5th (approx.) November: Michael Thurgood of Broadway said: *"I had just driven over the Pound crossroads in Broadway travelling north along Pound Road, when I clearly saw a black panther cross the road about 200yds ahead by the cricket field. At that distance, had it been a domestic cat, it would have been almost too small to see, but this was larger than a Labrador or German shepherd, and was clearly a cat from its shape and long tail."* **(Source: *This is the West Country*).**

Early November: Hinton St George man Dave Pitman said he had found, in woods near his home, a fresh carcass of a deer where the animal had been so badly mauled it must have been the work of a wild cat. **(Source: *Chard and Ilminster News*).**

November: A large black spotted near Dobvatt. **(Source: *Yeovil Express*).**

9th November: Christine Sunter of Donyatt Hill reported that, at around 9am, she was looking out on to fields by her home when she thought she saw a cow grazing *"but when it turned I saw its long tail and there was no mistaking what* it was - a big black cat," she said. "I got a really good view of it - there is no doubt in my mind." **(Source: *This is the West Country*)**.

14th November: Doris and Philip Jennings of Pine Avenue, Glynswood, Chard, say they saw a big cat at around 10pm. Mrs Jennings was saying goodbye to her two daughters, who had parked in the driveway of her home, and *"the next thing I know this thing charged past me,"* she said. *"It was so big and black and fast. Three leaps and it had gone."* **(Source: *This is the West Country*)**.

18th (approx) November: Simon Lancaster found a mauled deer while walking his dogs on Herne Hill, Ilminster. He said: *"I was amazed at the severity of the kill. The animal which killed this deer must have been some serious meat eater and no ordinary wild animal."* **(Source: *Yeovil Express*)**.

19th November: Ruth Turner, landlady of the *Volunteer Inn* at Seavington St Michael, was driving on the Cart-gate link road out of Yeovil to join the A303, when she saw a large black "panther-like" animal crossing the road. **(Source: *Yeovil Express*)**.

25th November: Nikki Martin, while out walking her dog, came across a dead deer along the riverbank at East Lambrook, near South Petherton. She said: *"When I looked, I could see there were only the bones and skin left."* **(Source: *Yeovil Express*)**.

24th December: *"Last night, 24th December, a very reliable person was driving his car just past Tropiquaria heading towards Simonsbath when he saw two big cats in the road in front of him. He has studied photos of big cats and was able to identify them as being lynx. The cats jumped over the hedge at the side of the road as the gentleman approached. The gentleman has worked in the countryside with animals all his life, is very knowledgeable of native wildlife, and is a very reliable person.*

It was dark, about 7.30pm and as he came round a bend before reaching Minehead, he saw the two cats in front of him in the road, one jumped over a hedge at the side of the road and the other stayed still and looked at the car and when he got closer this cat then jumped over the hedge. He drives a Vauxhall Astra and the hedge was higher then his car.

He said that their pointed ears stood out as he was looking at them from behind." **(Source: *Christoper Johnston & Anthony Bevan*)**.

Staffordshire
10 reports

4th February: Joggers spotted a large ginger-coloured cat in Griff Quarry near

Nuneaton, "sunning itself". (**Source:** *Heartland Evening News*).

8th February: Dog walker, Derek Stringer, reports seeing a large black cat in the Manchester area of Nuneaton. (**Source:** *Heartland Evening News*).

18th February: *"Myself and my mother saw a very large cat on Sunday 18th February 2007 whilst visiting my father's grave at North Walsall Cemetery, Walsall, West Midlands.*

Just beyond the car park, there is a small woodland area. As soon as I realised what we were looking at we got into the car, pulled up right to the edge of the car park towards the big cat and watched it for approximately 4 to 5 minutes. The cat sat there and stared at us then got up and walked towards us, stood just a few ft away and stared at us again before it decided to leave. It was gone in a second, there was no fear of humans apparent in this cat's behaviour.

I reported this sighting to the local authority on Monday 19th February 2007, and was amazed to get a call back from them a few minutes later confirming that they knew there were pumas in Walsall.

They told me that this puma was more than likely from Cannock Chase a few miles away. I was then asked by them to notify the Inspectors at the RSPCA Barnshill, which I did. I am amazed that we, the local inhabitants, had no previous knowledge that there were dangerous cats in the area. Mothers with children living and playing in this area should be warned. I have three grandchildren living within walking distance of this sighting."

Second email

"The cat was about as big (definitely no smaller) than a Labrador, but its tail was much longer, and the head wider and larger. Its body was very honed and powerful looking, it was completely black, and I saw no other markings. The cat was a mature cat, but I sensed from the look of it that it was a young, strong adult?

The first thing I noticed was the glassy stare, as though the eyes were reflecting the light as cat's eyes do. (My mother has since mentioned the eyes to me and shivers when she does).

The cat was already staring at us when I first noticed it. I had the feeling that it had been watching us long before we saw it.

It sat staring at us for quite a few minutes before it stood up walked a little way towards us (still amongst the leafless trees (therefore very visible) then it stopped arched it's back and relieved itself. We got into my car and I drove towards it very slowly, narrowing the distance between us to just a few feet. I stopped the car and my mother, myself and the cat just watched each other.

I felt no fear (I was sitting in the car). I think that I did not really believe what I was looking at. I tooted my horn on three separate occasions expecting to frighten the cat into running away, but instead it turned its head and just stared at me. I told my mother that I thought it was a panther, but felt, after putting it into words, that maybe I was being a little dramatic (this is England - how could it be a panther?)

The whole experience lasted several minutes, it was not a fleeting glance by any means and would you just know it, for the first time ever when at my father's grave I had not got my camera phone with me. This is unfortunate because I had plenty of time to have taken some excellent photos - the cat was practically posing for me.

It was the man from Walsall Environmental Health that told me it was a puma, after I described the cat to him. He also told me about the known pumas on Cannock Chase and said that this would be where the puma was from as this cemetery, quote: 'is on their run from Cannock Chase'. He also told me that there are known pumas on Barr Beacon (a well known country site area of Walsall)." **(Source: *Nigel Spencer*).**

February: "A witness contacted me yesterday regarding a possible sighting in Cheadle, Staffordshire. He is employed by three farms to control vermin levels, and an area he visits - an abandoned rail line - has seen some events he is unused to. He thinks he saw the hind quarters of a black cat several weeks ago, and has found 6 rabbits cleanly dissected at the centre of their body." **(Source: *D S*).**

February: "I was playing golf earlier this year at Perton Park, near Wolverhampton, and some friends and I witnessed a dead rabbit that had been skilfully devoured. It was quite muddy at the time, and we saw several partial large footprints near the kill. They were not clear enough to take a print as other players had trodden on them, but a friend who saw them thought they were not a dog's as he has a Rottweiller and said they were bigger than that." **(Source: *Martin Rainer BCIB*).**

26th February: "Sir, - On Monday at Trentham Gardens I saw a black panther. This superb cat was no more than 20-30 m from me and was, in fact, some 50 m from the deer herd, which I suspect it was stalking. The sun was strong and I had a clear and uninterrupted view of its form. It was of interest to me that another member of the public from Tunstall had only that day had a letter published on the same subject and this is why I have advised you." **(Source: *Stoke Evening Sentinel*).**

15th May: Sighting by Vicky Biddle, near Rugeley on Blithbury Road. The cat was black and had a very long tail. It was 100yds in front of her and was 3ft high. The eyes were a kind of blue in colour.

She said: *"I was driving, and the cat crossed the road in front of my car. The cat just went into the fields on the side of the road. I could see an outline and the tail. Also the eyes. My friends that live up the road think they have seen the cat in the fields at the back off their farm house."* **(Source: BCIB).**

10th July: A large black cat, 6ft in length, was seen at 7pm at the Hawks Green Island. This is some distance from Cannock Chase, in a highly populated area. **(Source:** *Martin Rainer BCIB***).**

15th July: A large black cat was spotted along Red Lion Lane in Norton Canes. This area is in a green belt that connects Cannock Chase with Chasewater. **(Source:** *Martin Rainer BCIB***).**

Early August: Sighting by three witnesses, at Churchbridge, Cannock on the A5 close to the M6 toll pay station. Its tail was probably as long as the body - thick and getting thicker towards the end. The cat was 2ft high, and approximately 4ft long, and was seen for 30-45 seconds approximately, eventually about 30 yds.

One witness said: *"I was driving along the A5 away from Cannock towards Brownhills in the early hours of the morning not far from Red Lion Lane. As I approached a lit up part of the road something walked out of the hedgerow into the carriageway and stopped.*

At first, I thought it was a deer. It stayed in the road, looking towards us, and as we got closer it could be seen to be jet-black with a very long thick tail. As we neared it, it casually turned round and went back the way it came from towards the toll road.

Several years ago my brother, and a group of fellow walkers (some of whom had cameras, but no one thought to use them) saw a large black cat clamber over a dry-stone wall dislodging some stones, in the Dufton area of Cumbria. Probably releases and they should be left alone." **(Source: BCIB).**

Suffolk
7 reports

14th January: Sighting at 10.30am (TM 292 861 GB Grid - OS Grid Number). The cat was black and had a very long thick tail - around 5 inches thick - that hung down and curled at the end. Height was around 3ft and its length (excluding long tail) was 4 – 5ft..

The witness said: *"I was travelling along the Flixton Road from Homersfield towards Bungay (O/S grid number TM 292 861 GB Grid) at 10.30am on Sunday 14 January. The weather was bright and sunny. Near the entrance to Readicrete pit, a car approaching me slowed down, I looked in my rear view mirror and saw a large black cat crossing from the pits to the woods. The other driver mouthed to me 'what the hell is it' but as the road was not safe to stop we had to move on.*

The cat stood approximately 3ft high and was around 4ft + in length (excluding the huge tail). The tail was very think and long - I would say around 5 inches in width - and hung down with a curl at the end. The cat was black and walked gracefully and slowly.

On investigating I would say it was definitely a black panther." **(Source: BCIB).**

February: Sighting of a cat that was a light (retriever) and had a long tail that looped down, and then up towards the end.

It was seen for approximately15 seconds at a distance of 150ft. It was a little larger than a retriever, but possibly longer.

The witness said: *"I switched on a security light to check operation before bedtime. A cat (?) 'loped' through the foliage at the end of the driveway at the bottom of the garden, but I lost sight quickly.*

Puma or lynx was my immediate thought. I was locking up for the night. We have just identified [27.06.07] some very large footprints. At 3in across they look big for a dog (certainly there are no dogs this large in the area) but the claws show." **(Source: BCIB).**

14th April: *"I've had a report from three women who were walking their dog in the Santon Downham woods in Suffolk. They said they heard hideous screams, which sounded like a cat and a dog fighting.*

I was over there all day yesterday, but found and heard nothing. The next day she reported that a muntjac deer was picked clean, near Kings Lynn. Not much left of it now – just the head and one front leg. She is going to call me if she sees anything else." **(Source: Terry Dye BCIB).**

10th May: Bartham near Bartham Hall? Witness saw what he believes to be a large black cat on the outskirts of Bartham at 06.55hrs.

He first thought it was a black bin liner blowing in the wind, until he realised that it was an animal and "definitely feline." He took some video. He said: *"The cat's tail was the thickness of a mans arm, long and sweeping."*

He did enquire at the Hall if they had any 'cats', but they replied in the negative.

(Source: *BCIB*).

11th June: Sighting by Mrs Brenda Sore of a cat along Mill Road, Peasenhall near Saxmundham at 20.30hrs. The weather was cloudy and clear. The cat was black, but she was too afraid to get too close to see any markings. She tried to take a mobile 'phone picture, but she not close enough. It had pointed ears. She is not sure how long the tail was but said that it came down at an angle, and turned a bit, rounded towards the end. She saw it for around 3 or 4 minutes, at a distance of about 100yds. It was about 2ft 6ins high and 3ft long.

Mrs Sore said: *"I was out for a walk, and as I came to the top of a big hill I saw it walking along the road in front of me. It seemed to hear me, stopped, turned round, and just stood still looking in my direction.*

I was quite scared that it may come back towards me. I stood still, as did it, so I thought I would have to go back again. When I turned to look back, it must have gone into the hedge so I carried on my way, the way I was going in the first place. I didn't see it again." **(Source: *BCIB*).**

17th June: Paul Newman was taking his dog, Coco, for a walk along Park

Lane, Kirton, at around 9.25pm when he saw the head and shoulders of a large cat emerging from bushes ahead of him. **(Source: *Suffolk Evening Star*).**

5th August: The black cat has been sighted yet again on Foxhall Heath in Ipswich .Cindy Lucas, who lives just off Foxhall Road, was left shaken after she spotted the huge animal as she walked her dog on Sunday morning. **(Source: *Evening Star*).**

Surrey
7 reports

29th April: Sighting at 6am. The witness said: *"I would rather not say exact location. I don't want to run the risk of it being shot etc. It was brown milk chocolate in colour, perhaps slightly ginger. It had rounded ears, but I couldn't see the tail as it was in a bush, but I could see its face close up, its chest area, and the size and length of its body etc. It was a very quick sighting, but long enough to see the face clearly. It was 3-4ft in height, and 4-5ft long - larger than a chunky male Labrador.*

I spotted a fox cub lying on the ground. I went over to have a look at it, and it seemed really strange, it had puncture wounds and blood on its neck/back of head and looked as though it had just been killed. I saw some movement in the nearby bushes, so I went closer to look, and stopped in shock as I saw its face - from where it was crouching/hiding in a bush, looking right at me! Then it turned and went into undergrowth.

I am 100% sure that it was a big cat. It looked exactly like a brown version of the black panther - my first thought was that it looked more like a bear's face, or a teddy bear's, face on a big cat's body!

It was brown milk chocolate colour, even slightly ginger, its face was very big and flat, it looked just like a teddy bear's face! Its nose was flat and bear/piggy like and had smallish eyes and small round ears. Its body was dumpy looking and low to the ground, it was about the size of a large male Labrador, a touch bigger though, and with a thicker, dumpy and heavy body.

I couldn't see its tail as it was in the bush, but I could see the length and size of its body as it turned

My wife is actually a vet nurse and animal handler, and I spend time walking dogs etc with her - I know that this wasn't some kind of dog - I recognised it as some kind of big cat, without a doubt.

I told my wife when I got home, and from what I told her she also was 100%

sure I had seen either a darker coloured puma, or a browner, less pigmented black panther/leopard, and from looking at photos online I am sure - as the panther bear-like piggy face in all the photos of panthers was what I saw.

From looking at reports online in the Surrey area, I was shocked to find reports of big cats right in this area - all the sightings of the 'Surrey puma' are online, but most of them are all a long time ago so I guess it couldn't be the same cat! I have decided to remain anon, and not give the exact location of the sighting, as I don't want someone to shoot it. It looked as though it had just killed a fox cub. There were large puncture wounds to the neck and back of head on the fox cub." **(Source: *BCIB*).**

26th May: Sighting at 2.00am in Byfleet of a jet black cat, with pointed ears. It had quite a long tail and the fur was wonderfully sleek, smooth and shiny. It was watched for about half hour, and it came just below the witness' window at one point. It was treble the size of any fox the witness had seen, and was like a very big dog. The witness had watched him come from quite a way down my road, as they had been looking out of the window enjoying the lovely cool air (don't sleep too well).

Follow up eMail.

"The reason I thought it was a puma... I described it to a neighbour and she said that was what it was. When I was talking to my daughter-in-law online that evening I told her, and she sent a picture on MSN of a puma, and it was exactly like the picture size, and shape, like a domestic cat but much, much larger -, like a Labrador fully grown I'd say. No sound whatsoever from it, and it was in the stalking position all the time I watched it, which was at least 20 minutes.

As regards the colour, next time my daughter and granddaughter were here, I made her stand under the outside light when it was completely dark in both black jacket and a brown one, and must say I am still 99 % sure it was black, although I must say sometimes things look a different colour when its dark.

My family are still thinking of sending me to the funny farm, and only half believe what I saw! So if you don't hear from me again you'll know where I am." **(Source: *BCIB*).**

25th July: A large black cat spotted by personnel on grass between the Keogh Barracks and a pub, Ash Vale Village. **(Source: *BCIB*).**

1st September: A Cheam resident, who asked to remain anonymous, said she saw a large black cat, resembling a panther, dart across her garden in Fieldsend Road two weeks ago.
(Source: *Sutton Guardian*).

3rd September: Darren Mason of Carshalton, was about to take in some evening hedgehog spotting with his family when a mysterious black cat darted across his lawn at roughly 8.30pm. (**Source:** *Your Local Guardian*).

1st October: Two witnesses at Abinger, Dorking Leith Hill saw, in the garden, a black cat, the size of which was smaller than an Alsatian, with a long tail. They were in the lounge when their dog started barking, so they looked outside. The animal made the noise of a howling cat, and the dog started barking. (**Source:** *BCIB*).

November approx: Witness spotted a large black cat near Croydon Airport "mooching through the grass" at Roundshaw Downs, Croydon Airport. (**Source:** *Evening Standard*).

Sussex
10 reports

8th January: A Langney man spotted a large black cat 'the size of a Labrador' near his home. (**Source:** *Eastbourne Today*).

c16th March: Joanne Walsh spotted a "black panther" in field near Gatwick Airport while she was travelling on a train from Brighton to Croydon. (**Source:** *The Argus*).

5th May: *"My grandparents were driving round an area near where I live called Coombes (near a place called Shoreham-by-sea along the south coast in the UK, England, google it!) at the weekend, so this would have either been 4th or 5th May. Coombes is basically a winding road through countryside, with a few houses and fields dotted here and there along the road. My Grandad went round a corner, and as he did so saw a 'big cat about a foot or two longer than a domestic cat' and was 'sandy coloured', in his rear view mirror. It sounds similar to a Scottish wildcat to me, but being as far away from Scotland as its probably possible to be in the UK I doubt it could have been one. Any ideas what this could have been?"* (**Source:** *Cactus Flapjack - CFZ.Com*).

16th May: A cat was seen by Kim Williams along Hindleap Lane at Ashdown Forest at 9.30pm during light rain. It was black with a long curled up tail, and was seen for about one minute at a distance of 20m. It was the size of a Labrador.

The witness said: *"I was driving along the road at approx 40 miles an hour - I never drive too fast as there are a lot of deer on this part of the road. I saw something black in the road - it moved slowly and it did not turn to look at me. It then jumped to the side of the road. I stopped my car and could see it in the full beam. It just moved off slowly and disappeared into the undergrowth.*

It was almost as if the cat didn't see me, it was totally not bothered by me. I think it may have been stalking something." **(Source: *BCIB*).**

21st May: Sighting of a brown/stripe cat with pointed ears, and a long curly tail, from a distance of 25 yds.

The witness said: *"I was driving back from Chichester, West Sussex, when the cat ran across the road in front of my daughter's car, near the fishing lakes. The animal was fast and was bigger than a dog with a brown stripe and long curly tail. It ran into undergrowth and woodland - they are still talking about it even now."* **(Source: *BCIB*).**

25th June: A passenger on a train reported seeing a lioness from the train at Wivelsfield railway station to the RSPCA. Simon Hancock believes his dog 'Hooch' was responsible for the sighting. **(Source: Mid-Sussex Times).**

Early November: A "family of big cats" seen wandering around the council refuse tip off Bexhill Road. **(Source: *Hastings & St Leonards Observer*).**

7th December: Sighting at 13.45 hrs at Telscombe Cliffs, East Sussex. The witness saw a black, leopard-like animal standing about 1.5ft tall and being 2ft long. Its tail was thick and long, between 1 – 2 ft.

The witness said: *"I was reading a book, and I looked up and saw it in the garden. I thought it was a stray dog at first, but then it moved and I knew it was a cat. I saw it take three steps, and then stop with one paw raised looking at a tree near my garden fence. It was far too big and thickset to be a domestic cat. Out of shock and excitement, I jumped up, opened the sliding door and ran out to get a better look. I wish I had stayed still to watch it, but I think that I was in mild shock because I could not believe what I was seeing. It shot off faster then I have ever known a domestic cat to move, I think it went through a gap in the fence at the back of the garden."* **(Source: *BCIB*).**

13th December: Sighting at 7.20am at Westfield, near Hastings, East Sussex, when the witness, who wants to remain anonymous, saw a black cat, which was larger than a German shepherd dog. It had no other markings and a long tail.

She writes: *"It was a clear frosty morning, and the cat was lying down. When it noticed me, it stood and then ran across the field towards the woods. Very sleek and fast. Looked at my dog, and my dog (a West Highland terrier) growled and stepped back.*

This is the first sighting I've had, although a fellow dog walker says she previously noticed something similar. I would like the cats not to be hurt" **(Source: *BCIB*).**

25th December: Robert Gillespie from Brighton said that he was on the edge of the south downs when: *"I saw a huge black cat, no markings that I could see, he was black as night. Pointed ears. Long black tail about 1m long, its height was 50-70cm and length - 110-15cm. It was either a puma or a black leopard. I was out snaring rabbits. I finished setting a snare and when I stood up, I saw the bastard trotting alongside the outside of a small wood and then he was gone."* **(Source: BCIB)**

Teesside
4 reports

March: Sighting of prints by Laura Henry at Wynyard Woodland Park at Pickards meadow (west side near main pond), Stockton. There was lying snow with light over-casting. The prints were large at approximately 3-4in across with no claw marks. They were definitely not made by a dog, a fox, a badger or any other large native mammal.

The prints were a cat print but extremely large - too large for domestic cat. The track was approx 5-6m long, and she followed it from [...] the woodland, across a snow covered grass area and back into woodland. It had snowed earlier in the day so the prints were no more than 3 hours old. There have been several sightings of a large black cat like animal in the same field and surrounding woodland, and dead roe deer and sheep have been found. **(Source: BCIB).**

June: A sighting of a large black cat in Billingham. No other information is known. Apparently, it was featured on the Richard & Judy show. **(Source: Ian Bond).**

June: *"I got a call from a former colleague on Stockton Council who manages Cowpen Bewley Woodland Park. The drive-on lawn-mower man was cutting the narrow grass paths through the site this afternoon when a bbc jumped out in front of him. It was described as black, knee to thigh height with rippling muscles. The lawn-mower man does a bit of shooting so is not a complete novice to wildlife.*

I actually went and investigated this one as the sighting was only about 90 minutes old by the time I got there. Unfortunately, there wasn't anywhere where a sign would be obvious. CBWP is pretty big - the wardens used to brag that the site in its entirety was bigger than the City part of London. It's mostly 15-year-old woodland plantation, which is very dense as it has never been properly thinned, but there are some rank meadow areas as well. No mature trees for scratching or dragging ungulates up, and no areas of mud for prints." **(Source: Ian Bond).**

December: Sighting of two cats at Wolviston near A689. The witness sees them one a large black panther walking along track alongside his house, and then, more frequently, a medium sized cat-carrying rabbits along the same track. (Witness acquired photos of this second animal, which appears to be a large feral domestic). **(Source: *Chris Hall*)**.

Tyne & Wear
1 report

13th January: *"I saw a fawn coloured cat crossing the A1 near Kingston Park Airport turning (Newcastle. It was night time (10.20hrs) so I am not too sure about the colour.*

From my rear view mirror it appeared to be bushy. As for length, it was in proportion to its body, which was far bigger than a pet cat. When I first saw the cat, it was running along the edge of the road. It jumped back and hid/cowered until my car had passed. It was a lot bigger than an average cat (and I am an owner of a rather large cat). When I initially saw it I thought of a lion cub due to its size. Because I thought this strange, I kept my eyes on my rear view mirror to see what it did next. It immediately bounded across the A1. It was too big for a domestic cat so I looked at its tail to see if it could be a fox, however its movement was definitely that of a cat, pounding two feet at a time." **(Source: *BCIB*)**.

Warwickshire
2 reports

28th May: Sighting by William Guild. It was late afternoon, in a field near Coombe Abbey, just by the A46. The cat was jet black, with pointed ears and a long thick tail, and it was seen for around a minute at a distance of a 100m. He was not sure, but thought it was around 5ft in length - the size of a fully-grown Labrador.

Mr. Gould said: *"We were walking towards the end of the field, and we saw this large black cat-like animal rise from the grass. We initially thought it was a black dog, but when it ran towards the hedge, we noticed it was too sleek to be a dog, and the long tail gave it away as it was more flexible like a cat's tail. We ran to the end of the field to see if we could catch another glimpse of it, but the animal was gone and nobody else was around. We are assuming it may have been somebody's dog."* **(Source: *BCIB*)**.

16th December: Sighting in Coventry at 10 pm of a black cat which was seen for 40 minutes on and off.

The witness said: *"When metal detecting, my friend came and told me to keep an eye on the top of hill as it seemed to be watching me from about 300ft. It was not there at that time, but when I got to the end of the field by a pond there was a very strong smell like cat pooh. We thought one of us had trodden in it, but our shoes were clean. On the way back, I spotted it on the brow of the hill. I looked round to see where my friend was to tell him, and when I turned back it had gone. When I looked back it was coming down a hedge towards me, and its back caught a branch sticking out of the hedge. I went to look afterwards, and the branch was at least 2½ft off the ground. It turned and, as it went along the hedge, I couldn't see it as it blended into the hedge. It was only visible as a silhouette with the moon behind it. I heard a sound earlier like a kitten crying."* **(Source: BCIB)**.

Wiltshire
31 reports

January 2007: At Savage Cat Farm, Gillingham a black panther and caracal lynx were reported by Mrs. Hall. **(Source: *Marcus Mathews*)**.

8th February: *"The witness was walking his dog at 14.50hrs on a remote part of Salisbury Plain, when he saw a large cat on a tank track from the corner of his eye 200yds away. 'What the hell is that?' he thought to himself. The animal was 'cat shaped, dark brown or black, and stood sideways' looking at the witness. He walked slowly towards it for 10yds - the animal did not move. It was stood in a 3ft track, which it covered, and he estimates the animal was about 22in at the shoulder. He took another three paces forward, but the animal did not move, just stared.*

The witness decided to retreat and walked back to his 4x4 and then headed back to the area. He pulled up at the spot, but could not see the animal, then from the opposite side he managed to catch a glimpse of the rear of the cat and its tail disappearing into a plantation of trees. He did think about going in on foot to look for it, but decided against that.

He reported the incident to the rangers and the police. The police said that they knew all about the cats that lived in that area, and that there was a colony of them! The witness returned the following day, but found no signs." **(Source: BCIB)**.

3rd March: Stuart White, the landlord of *The Bell* public house at Chalfield, Westbury, was travelling through Wingfield when he spotted a dark, feline shape.

Mr. White said: *"I saw what looked like a panther out of the corner of my eye. It was pure black and about Labrador size."*

He stopped the car and watched the cat as it stalked across the field. It was only about 30 yds away. He continued: *"It just ignored me and appeared to be carrying a cub in its mouth. I haven't ever disbelieved people when they say they have seen large black cats, but this was definitely larger than a large cat."* **(Source: *Marcus Mathews*).**

March: At Stourhead, Wiltshire a black panther was reported by a gamekeeper. **(Source: *Marcus Matthews*).**

9th March: Sighting near the A4, between Box and Corsham, Wiltshire at 0700hrs.

The witness said: *"Looking from the edge of a small wooded area across a farmer's field with the sun behind me, I sighted a cat-like animal walking along a farmer's track towards another small wooded area. The animal was black, and roughly 1.5m long and 70 – 80cm in height. Its movements were not like a dog's, and there was no one else in the area."* **(Source: *Chris Mullins Beastwatch UK*).**

March: On Salisbury Plain, Kevin Matthews saw a black panther. **(Source: *Marcus Matthews*).**

21st March: 'Is this a black panther?', asked the *Wiltshire Times*. Matthew Finch, of Blease Close, was shocked to see the animal in a field behind Airsprung Beds in Canal Road, Trowbridge, as he left for work at 7.30am. He said:

"As I walked out the door I thought 'hang on a minute'. It looked like a cat, but it was a lot bigger. It was definitely a cat. It walked like one, and had its big black tail in the air." **(Source: *Wiltshire Times*).**

25th March: Two policemen, and a civilian officer, were shocked to see a big black cat run across fields in broad daylight on Easter Sunday. **(Source: *Wiltshire Times*).**

15th April: Taxi driver Nigel Davis, of Woods Lane, Chippenham, had a late night encounter with what is locally known as the 'Beast of Biddestone'. **(Source: *Wiltshire Times*).**

4th May: Sighting of large back cat in Canal Road, Trowbridge - see photo on *Wiltshire Times* website. Also sighting at Bishopstone again. **(Source: *Marcus Matthews BCIB*).**

8th May: Sighting by Julia and Danny Rea and two other witnesses at 8.45pm, in Warminster, in woods close to Battlesbury Hill. It was black with no other markings.

One of the witnesses said: *"I didn't notice tail - it wasn't up! It was seen for about 2-3 seconds at a distance of 50ft. Its height was about 20in - the cat was stooped and 20in is the height I would estimate when stooped - not its full height. The length was about 30in. It was seen briefly – we first realised something was up there on Monday 30th April.*

My husband was walking the dog when he heard what he described as a duck or crow being strangled. He looked but saw nothing. A few minutes later, whilst looking into a field for rabbits, he heard something come running towards him through the bushes very fast. He said it was, by the sound of the breaking branches and foliage, something large, but didn't stop around to wait for whatever it was. He ran to a clear spot and threw stones into the spot to frighten it away. He then made a hasty retreat home.

A few days later, when on the same route with the dog at 8.30pm – ish, we heard breaking twigs and branches again and turned round and came home - this happened the following evening at the same time.

Last night, Tuesday 8th May, we went up with my sister, my daughter, my husband and me. My sister heard moaning, while my hubby and I were looking into the rabbit field. She called us back, but I heard nothing. My husband then heard a noise, but we saw nothing. As we decided to go home, my sister called me about another noise and as I turned to her I saw a large black cat run out of the trees across the pathway and into the trees the other side, about 30ft behind her, and about 50ft away from me. The cat was stooped as if to keep a low profile across the pathway. His head was level with his back (in a straight line) - I can't say I noticed a tail as I was shocked to see a large cat, more expected to see a wild boar! My daughter, 12-years old, also saw it and commented 'awwww a pussycat!' I, however, said something a little different. We made our way home the long way seeing as it crossed our pathway home!

My sister took a photo of a rather large paw print at the bottom of Battlebury Hill with her camera phone during last week. We cannot be sure if it is a cat or large dog print, so it is possible it is nothing, but we have the photo still on our phones. We also took a photo of the four paw prints of how it was stood if that's any help."

Wiltshire investigator Marcus Matthews reports:

"I telephoned the people last night. Panther was seen in a copse behind Bishopstrow Court Hotel, Warminster near Battlesbury Hill. It growled and ran at them through undergrowth. It was 20in high and 30in long and jet black, with no visible tail - probably as it was seen head on. I will go up there and look. They are sending photo of paw-print taken on mobile phone." **(Source: BCIB).**

May: Panther seen at Berwick St. John by the chef from the *Talbot*. **(Source: *Marcus Matthews*).**

May: Young man spotted a panther near the *Talbot*, Berwick St. John. **(Source: *Marcus Matthews*).**

May: Panther seen near Windgreen Hill, Berwick St. John. **(Source: *Marcus Matthews*).**

May: Panther seen by Mr Follett, of Berwick St. John, crossing road near Ferne Park. **(Source: *Marcus Matthews*).**

May: A large black cat spotted near Shaftsbury. **(Source: *Marcus Matthews*).**

May: A large black cat was spotted at Ludwell on scrubland hillside, near the *Red Lion* pub. Fifteen lambs were killed on the farm belonging to Mr. Richard Pocock. This was reported in the Blackmore Vale Magazine. Marcus Matthews and Jonathan McGowan investigated. Signs of attack by small dog with powerful jaw e.g. Staffordshire bull terrier - Jonathan McGowan skinned the sheep. There were signs of badger digging near-by. **(Source: *Marcus Matthews*).**

2nd June: At 5pm, Marcus Matthews saw a large black cat-like animal similar in size to greyhound with the sun behind it, crossing road below Batts Lane, Potter's Hill, Crockerton. **(Source: *Marcus Matthews*).**

8th June: At 8am, Marcus Matthews saw a large tortoise-shell cat - larger than domestic cat – with a long tail and long body crossing the road at the bottom of Batts Lane, Potter's Hill, Crockerton. It was the size of a fox. Was it a feral cat, a chaussie, or a puma cub? **(Source: *Marcus Matthews*).**

11th June: *"Spotted large tabby cat in field in Crockerton, which was larger than a domestic. Probably a local feral, but certainly the size of a fox."* **(Source: *Marcus Mathews BCIB*).**

15th June: Fourteen-year-old Brogan Moulden spotted the cub of a cat on the roadside in Castle Eaton. The animal had been killed by a car, but residents in the village were amazed to discover a wild cat had been roaming their streets.

Rick Minter reports: *"Frank Tunbridge 'phoned the pub where the dad works (Father of the 14-year-old lad who found the dead cub). Apparently, it's a Bengal cross. Following the publicity they've had several calls, and three people have claimed the dead cub as their own."* **(Source: *Swindon Advertiser & Rick Minter*).**

June: There was a sighting of a black panther in a copse behind Bishopstrow Court Hotel, Warminster, by Mr. and Mrs. Danny, and Julia Rae of The Deanlands, Warminster. The cat growled and rushed at Mr. Rae through bushes. It was seen crossing the field by Mrs. Rae.

Marcus and Heather Matthews investigated - a roe deer was seen, and possible claw marks on a tree. **(Source: *Marcus Matthews*).**

July: There was a sighting near Marlborough by a 15-year-old, of a panther in a crop circle. **(Source: *Marcus Matthews*).**

August: *"I'm 15 years old, and I know you can't believe a 15-year-old, but I was at a site next to Sugar Hill in Oldborn, Aldbourne, looking at a crop circle. On the way back down the hill, my friend stopped and asked 'what the hell is that?' I looked, and about 100ft away was a black panther. It was the size of a German shepherd dog, but bigger in the belly, with a long tail. It then moved up towards the top of the field, but stopped in middle wagging its tail. After seeing that we ripped an iron bar from the ground and made our way back to the car, which was in the direction of the cat. I can't remember the last time I've been that scared."* **(Source: *Quincy00 / Weird Wiltshire Forum - forwarded by Katy Jordan of BCIB*).**

29th September: A two witness sighting at 18.00hrs in Horninsham, on the road between there and A362 (Warminster). It was a solid charcoal grey (not quite black) with pointed ears. Its tail was noticeably long in relation to the body, and curved upwards, with an even diameter along its length. It was hard to be sure, but it was estimated at twice the size of a 'normal' cat – a large unusual feral.

Tom Smith said: *"I stopped car when I noticed a large cat in a field adjacent to the road; the field sloped steeply upwards away from the car (the cat was approx 60yds from us).*

The 'cat' was walking up the slope quite slowly, and stopped when it heard my car. It turned its head, looked at us over its shoulder for 10-15 seconds, and then continued up the hill and into the undergrowth at top of hill. The very confident (arrogant?) way it was walking seemed very unusual. I suppose that there must be a number of escaped big cats that have established themselves in the countryside. There are huge areas of forest (Longleat etc) in this area, with plenty of places for them to live/hide. And, no I don't think anything should be done about them." **(Source: *BCIB*).**

1st October: A worried father has called for action by the authorities after he and his family saw a mysterious, big black cat prowl around the edge of a school playing field for 10 minutes in Malmesbury. **(Source: *Western Morning News*).**

15th October: A large black cat was spotted on a farm near Hullavington. **(Source: *Wilts and Gloucestershire Standard*).**

October: A black panther was photographed near Dyres Garage, near Hensridge, in the Blackmore Vale country. **(Source: *Marcus Matthews*).**

25th October: A large black cat was spotted on the outskirts of Malmesbury. Local residents believe it lives in field on the edges of Malmesbury. **(Source: *Wilts and Gloucestershire Standard*).**

28th October: Oaksey resident David Reeve saw a big black cat in a field behind his garden at 8am. **(Source: *Wilts and Gloucestershire Standard*).**

5th November 2007: Sighting by Esther Stone at Chestnut Barn, Cold Ashton nr Chippenham (OS) SN14 8JT. It was the size of a Labrador dog, and was seen for around 2 minutes, at a distance of approximately 30m away.

Ms. Stone said: *"I saw a large black cat in my garden, from a high up window looking down into my garden.*

This was about 3 weeks ago. I looked again as I thought it might be a dog. It was pure black, and I couldn't get over the size of it. I didn't think anything of it again until my neighbours talked about missing chickens and sheep in the field right behind my house.

Tonight, my partner saw bright green eyes at the bottom of our garden and he is convinced it wasn't a house cat because it also looks so big.

This cat, or whatever it may be, is very large and is coming into our garden. I have three very young children, and am starting to get very worried. How do I go about getting some cameras set up to catch it on camera? A lot of strange things have been happening over the past few weeks, which we have only just found out about. This cat, we think, is in the field behind and seems to be getting tame, and doesn't worry about coming near to our house. Also a neighbour across the road has done plaster casts of footprints and they have been confirmed as puma prints - this came from Bristol Zoo. Another neighbour has seen this about 2 years ago." **(Source: BCIB).**

Worcestershire
3 reports

23rd May: Sighting by a lamper at Inkberrow in Morton Underhill Woodland. He has been a lamper for 15 years, and said he gets used to the animals at night - their eye-shine, height etc.

He said: *"I saw, 300yds, away eyes which I could not readily distinguish; I was using a rabbit squealer at the time. I moved closer, then the 'eyes' ran, in under 30 seconds, 40yds. They stopped and stared back, again, at me. I moved forward again, and the pattern was repeated. The eyes were about 3½ft off the ground - wider apart than a fox's would have been."*

The next day, in the area, a sheep was discovered with its muzzle bitten off. The witness also found fur on a barbed wire fence at the spot he saw the 'eyes.' **(Source: BCIB).**

18th December: *"This morning at about 10:30am, my mother in-law and I sighted a large black cat the same size as my Golden Retriever, running along the edge of a hedge near Harvington, Evesham."* **(Source: BCIB).**

22nd December: *"My daughters and I were walking this afternoon and saw what I believe is a wild cat. It was running away from us from 10ft as I turned and noticed it. The usual, what's that? In a split second you assess all the possibilities – dog? Too feline, so no. Domestic cat? No, far too muscular and it had distinct black fur at the tips of its ears. Light brown, longer hair to its tummy with darker tail – deer? Sheep? All no. Suddenly it dawns, a wild cat? Surely not round here? But sure as eggs are eggs there's no other explanation for it. We saw a wild cat of some sort today in the woods behind our house, 40 minutes ago! Don't know who else to tell who'd believe us!*

The area involved was Dunhampton, in-between Droitwich and Kidderminster. It was the colour of the puma, but the size of a lynx. It had a dark bushy tail with white detail on it. My 7-year-old daughter said she'd seen one like it on 'Roar' on the BBC." **(Source: BCIB).**

Yorkshire
64 reports

Early January: A dog mauled to death in Edlington Woods, Doncaster, may have been the victim of a big cat, some say. Enquiries by Mark Fraser revealed that several weeks ago a large Doberman went missing in the woods, and has been spotted once or twice, and is now believed to be living wild; could this be the culprit? **(Source: *BCIB*).**

4th January: Leon Daines, 21, believes he may have stood within yards of a large mysterious feline; he was returning home with a friend through Edlington Wood in Doncaster, when the pair became distracted on Tofield Road.

He said: *"We were walking through the wood when we heard a very loud noise. Usually when you hear humans walking over sticks they might crack a little bit. But this made a much bigger crunching noise. We didn't know what it was, but it was walking up by the side of us in some brambles. My mate shouted 'squat down, squat down', but I started growling and the noise stopped. It lasted for about 10 seconds.*

I was very scared, we were both really panicky."

Mr. Daines, who takes a keen interest in wildlife, has not returned to the wood since the incident. A few years ago he caught a glimpse of two large cat's eyes while driving through the same area, but rubbished the idea of a 'big cat'. **(Source: *Doncaster Free Press & BCIB*).**

7th January: Sighting by Brian Poole at 11.00am in Russell Hall farm field at back of St Peters Court Whitby.

Mr. Poole said: *"I was working in my kitchen with my two friends when I noticed three deer come out of the woods looking startled. When we all stopped to look, they stopped, looked back at the woods, then took off rapidly down the field. We followed their movements down the field. I then noticed a dark (black) shape at the top of the field where the deer originally had been. When I shouted at my friends to look it must have heard me and it turned around and disappeared.*

The cat was about 2½ft high and 5ft long, and it had a long tail which curved upwards at the end." **(Source: *BCIB*).**

15th January 2007: Report of sighting by Sharon Hutchinson, who was walking to school at 08.15hrs with three of her friends when they saw a large black cat on the old railway viaduct near Caedmon School. She said it was just slowly walking along the path, had pointed ears, was about 5ft long and 3ft in height.

(Source: *BCIB*).

18th January: *"I was driving to work at approx 8.15am on the Driffield road to Scarborough. Just before leaving the village of Foxholes, I saw a black cat-like animal about the size of a Labrador. It crossed the road from right bank to left bank. There were two other cars in front of me."* (Source: *BCIB*).

28th January: *"I am writing to you as I saw what seemed to be a big black cat at 7.45pm. I live in the East Riding of Yorkshire, and I spotted the cat as it ran across the road whilst I was driving into the village of Burton Pidsea. I was with my girlfriend at the time, and we both remained quiet before asking one another 'Did you just see that as well'?*

It was a large figure that was the size of a black Labrador, but was more shaped as a big cat, and how it ran across the road. It was very dark, but we didn't see the distinctive 'eyes' that many people say you see staring at you, but - as I said - it ran in front of the car and was only visible for a couple of seconds." (Source: *BCIB*).

1st February: On Thursday night near Doncaster, at the edge of Eglinton Woods, a taxi driver was startled by a "mucky golden coloured cat as large as a Labrador" which came running out of the woods, across the road in front of him and away into woods on the other side.

Later that evening, a woman telephoned to say that she heard a "strange wailing" coming from the same woods which her back garden backs on to. This was about half a mile from the taxi driver's sighting. (Source: *BCIB*).

17th February: Sighting by Mr. Brian Barton at 1.45 p.m. on Bridlington Road, between Coniston and Long Riston. Black colouration with a long tail, approximately 2ft - 3ft. it was seen for ten minutes at a distance of 40m and was approximately 2ft in height and 2ft - 3ft in length.

Mr. Barton said: *"Cat was running across the field chasing something. I was a passenger in the car with direct view of cat, which was also seen by my daughter, and wife, who was driving. It was very clear to see."* (Source: *BCIB*).

11th March: Day-tripper David Sykes said he saw a black panther between Beckermonds and Deepdale, in Langstrothdale, between Hawes and Buckden. (Source: *The Northern Echo*).

March: Sighting at 11.30pm in Batley. The witness said: *"It was black, with no markings that I noticed. It had rounded ears, and a long tail, curled up slightly at the end. It jumped over a 5ft wall easily - I'd say it was about 4.5/5ft long, not including tail. It was seen only for a matter of seconds. I was in a car, about 30ft away (I'm a bad judge of distances, I'd say around 6 car lengths away).*

I was driving home, it ran across the road in front of me, jumped a 5ft dry stonewall with ease, I stopped at where it had jumped, but was too scared to get out of my car. I've been interested in Forteana and Cryptozoology for a while, being an avid reader of the Fortean Times. *I'd be interested to know about any other sightings in the west Yorkshire region. I have discussed this with friends who have also told stories - recent stories of sightings in west Yorkshire. Could you give me information about sightings in my area?*

I believe that most cats have been introduced due to human negligence; pets that grew too big. They should be caught, but only for the care of themselves. I can't imagine it being very nice being a large cat wild in Britain. Re-homing at a zoo or sanctuary is the best place for them then." **(Source: *BCIB*).**

March: A motorist spotted a large black cat panther-like animal in fields on the outskirts of Osgoodby while driving on the A63. **(Source: *BCIB*).**

March: Seen several times by farmer at Glaisedale, North Yorkshire, a black feline, whilst he was out lamping **(Source: *Chris Hall*).**

March: Sighting at Whitby of black 'panther'. It was seen following deer near housing Estate on the south east edge of town. Also, around the same week, a very close sighting of a large black cat, where another witness investigated 'baby's crying noises' apparently in a tree, and then saw a large black cat come down the tree, and run off. **(Source: *Chris Hall*).**

24th March: Sighting at Wakefield at 5.15 p.m. at the Hazlewood Castle Hotel, Leeds. It was seen from the A64 A1/M road junction. It was early dusk on an overcast, cold and damp day. The witness said: *"It was jet-black - couldn't see any markings. It had rounded ears. The tail was very broad and 'furry' especially towards the end - quite blunt - probably over a foot long. Seen for only a few seconds, 100 to 150yds away, I think. It was about the same length as a large Alsatian (nose to tip of tail), but shorter in height.*

The cat was walking along the boundary of the field and a wood. It wasn't moving quickly. I was a passenger in a car on the A64/ A1/M junction. It was so striking - it just didn't look like a normal domestic or even feral cat - it seemed much too big." **(Source: *BCIB*).**

25th March: Newbald to S Cave - a drop down near an old railway bridge, near chalk quarry. The black cat walked with a slinky kind of movement - it's not a pussycat. Walking along the hedge line of a field. Geoff Featherstone is following this up and we are waiting for the report. **(Source: *BCIB*).**

27th March: *"I can report having seen a big black cat in Cloughton in a field near the railway line yesterday at 6pm. The cat was the size a small Labrador - black, long tail and long body. We both had a clear unobstructed view of it walk*

across the field to the railway line, and then back across the field to some bushes into which it disappeared."
(Source: *Ann Tindall*).

30th March: - see 27th. *"We've seen it again, same place and time. It's quite near to our house and was only yards from people walking on the railway line, but they did not see it. The time and date was 5.55 pm, 30th March. There used to be loads of rabbits in this particular field, however we've not seen any for a few weeks. I really want to get a photo, but haven't been lucky enough yet."* **(Source: *Ann Tindall*).**

3rd April: Sighting at Hedon. The lady rang several times whilst she was actually viewing a large cat and a smaller one in fields near her, behind her house. She rang the *Daily Mail* in desperation, we finally got hold of Geoff Featherstone who followed it up. Waiting for his report. Kids also said they saw the big black cat the day before. Neighbours saw it earlier in the day. **(Source: *BCIB*).**

15th April: Edlington Woods, Doncaster. Leon, and his girlfriend, went to the woods to look for any signs of the animal that gave him a fright on the 2nd of January. He shone his torch, which lit up a pair of eyes. He said: *"The eyes were in the bottom left hand corner of the field - a pair of big green eyes, really vibrant in colour! My girlfriend has better eyes than I do, and she saw them as well. After we saw them, we had to leave the woods, as she was too scared to walk any further! The good thing is that she now also believes that there's a cat in the woods!*

When we saw the eyes they were low to the ground looking upwards. They then lifted to around 3 - 4ft off the ground! I have also found some kind of scratch marks in the mud (photo is attached) what is your opinion on these mate? I have found out that there is an old railway line that was used for the transit of coal. It runs along the old pit and then towards Doncaster. There are no tracks there, but the line it ran and bridges are still there. This location is around half a mile from the woods."

Mark Martin met up with Leon and took a walk around the woods. While there, they checked the meat bait and Mark reports: *"Earlier today I was with Leon Daines, checking around the sand traps and bait he has laid in Edlington woods (there have been several sightings in this area and one night Leon saw a pair of large green eyes, 2½ft off the ground). We found the remains of a large piece of meat that Leon had tied to a tree, about 3ft off the ground. I was very interested, as the small piece remaining (underneath the strands of nylon rope that fastened it to the tree), looked as if it had been cut very cleanly, very neat, looking like it had been rasped. It was also covered in what looks like dried saliva. I think it may have been a big cat. It looks like the rasping action of a big cat tongue to me.*

Leon has collected the meat remains, together with some tree bark and nylon rope that was covered in saliva. He has them in plastic bags, in a plastic box.

The ground in the area was dry and hard, not showing any tracks.

The pictures don't really do justice to how clean and neatly the meat had been eaten away. There was also, what appeared to be a tooth mark in the meat, puncturing deep into it. The hole was about 9mm in diameter. I really was struck how clean the meat was, looked like it had been brushed." **(Source: BCIB).**

30th April: Sighting at 19.30hrs, near the North York outer ring road of sandy (puma colour) animal. [...] long uplifted toward the end. It was seen for less than 30 seconds, 50-70yds away. Its length was approximately 4ft. The witness does not know the height as it was walking in a ditch.

The witness said: *"I was sat near the York outer ring road, and just looked up, the cat was walking in a ditch with only its back visible. It just walked towards the field corner then disappeared into undergrowth. I didn't have time to get my camera phone before it was gone."* **(Source: BCIB).**

1st May: Trucker John Tordoff glanced into a field on a country lane near Sherburn, N Yorkshire, and saw what he believes was a black panther. **(Source: *The Sun*).**

10th May: Sighting at 4pm. A. Varley, and another witness, were travelling on the train from Doncaster to Retford when, about half way through the journey, the cat was calmly walking in the middle of a field. It was black with a tail that was very long and hung low then turned up. It was about 50yds away.

Mark Martin reports:
"The witnesses said the sighting took place halfway through a train journey between Doncaster and Retford. When you check this out on a map, it indicates the area around Bawtry, where Russel and Barbara Fearn had their sighting on 5th August 2006. I will have a good check round the area and report back if I find anything interesting." **(Source: *BCIB*).**

13th May: A couple saw what they described as a "big black cat – like a black leopard or panther" near the village of Westerdale. Now dubbed the 'Beast of the Bay' as it's frequently seen around the Robin Hood's Bay area. **(Source: *Whitby Gazette*).**

May: One family found themselves staring into the `shiny eyes' of a large black cat-like animal near a Cleckheaton playing field. **(Source: *The Press*).**

May: *"I saw your site on the internet, and found what you wrote very interesting, as I have seen a big cat myself. It was black and quite big, bigger then an Alsatian. It was while I was walking my dog over the moors in Morrends near Doncaster. To be honest I think that local authorities know that there are big cats out there and are just hiding it from the public, which is disgraceful."* **(Source: *Nigel Spencer RLPW*).**

May: *"I have seen these big cats several times in the same location. My daft terrier once chased one some 200m where the cat actually ended up a tree looking down on my dog. This was quite good as it gave me the size difference.*

On another occasion I was just 1m away from a big cat and what appeared to be its cub. I have kept the location very much to myself because you will always get the big white bounty hunter wanting to kill them.

Let me tell you that the animals in question live alongside various livestock without any reported problems, maybe because the location is a peat moor and there is lots of smaller natural prey. The picture in the Goole Times does not relate to the ones I have seen. I looked up various pictures and if my memory serves me right it was called a European lynx. By the way, the location is 6-8 miles from Drax." **(Source: *BCIB*).**

Late May: Female witness spotted a large black cat on the Drax Village common; brown, as big as a Labrador with a dark brown stripe along its back. She at first thought she was seeing things and did not report it to the authorities. **(Source: *BCIB*).**

10th June: A housewife in Church Lane, Dewsbury Moor, thought she was seeing things after spotting a large black cat near bushes in the grounds of Westmoor School.

She said: *"I couldn't believe my eyes. I thought people would think I was mad, and then I read the story in The Press."* **(Source: *The Press*).**

11th June: A cyclist spotted a large black cat – which he described as similar to a panther – on his way to work, on the Spen Valley Greenway near the Rohm and Haas chemical works off Heckmondwike, Cleckheaton. He said it was the size of a Labrador. The witness said he was scared and then decided to 'get the hell out of there'. **(Source: *The Press*).**

15th June: A big cat the size of a rottweiller dog has been seen roaming in Cleckheaton. A family out walking their three dogs saw the animal – which had pointed ears and a long tail – on a pathway in some woodland behind Whitcliffe Mount School. **(Source: *Spenborough Guardian*).**

June: Rosealene Ballan from the Spen Valley area, Cleckheaton, reports that her husband has seen a large panther-like cat in the area. **(Source: *The Press*).**

June: Jayce Allatt of Ravenshouse Road has been mystified by happenings at the semi-wild end of her back garden at Heckmondwike Road, Cleckheaton.

"Something big is coming in there," she said. *"It digs holes and climbs a tree. Clumps of leaves and such keep coming down. I keep filling up the holes but the next day they are scrabbled out again. I have lived here for more than 50 years and nothing like this has ever happened before. It doesn't scare me though."* She has found mangled takeaway cartons inexplicably left there. **(Source: *The Press*).**

Mid-June: Barby near Selby - Motorists on the A19 spotted a large panther-like cat walking across a nearby field. **(Source: *BCIB*).**

29th June: *"Hi, I was travelling home from work on Friday 29th June, going east on M62. When we got to the Goole fly-over both of my work colleagues noticed a very big cat stalking some sheep that were on the river bank. They both said it was creeping along a hedgerow. They say they thought is was a normal cat but a big one, but then realised how far away we were and how big it was. They had a very good view, as we were right at the top of the fly-over looking downwards to the riverbank they say. It was jet black with no marks anywhere and they said it was very long. Unfortunately, I didn't see it as I was driving (gutted)."* **(Source: *BCIB*).**

2nd July: Sighting at 18.15hrs on Cutsyke railway crossings, Leeds Road, Castleford, West Yorks. It was crossing the banking between the railway tracks and the allotments. The weather was humid, dry at the time, but all day on and off showers some heavy at times. The visibility very clear. The animal was dark grey, not unlike tabby cat markings. The tail seemed to be quite thick and as long as the body of the cat. It ended in a point with a very pronounced white tip. The tail was between 1-1½ m long and the cat was ½m high, and it was seen for 30 seconds – from 20m away.

The witness said: *"I was walking my terrier on the raised banking between the railway and the allotments. The cat appeared from the tree line to my left, and crossed to the trees on my right.*

My terrier chased the cat, but I believe he never found it - normally when he traps wildlife e.g. squirrels, hedgehogs etc., he screams and barks with excitement. I'm glad he didn't as when he gave chase I feared for him. The allotments contain lots of chickens, I spoke to one of the allotment holders who said that they had had quite a few of their chickens killed." **(Source: BCIB).**

2nd July: Telephone call after a piece in the *Pontefract and Castleford Gazette*: A large "grey cat-like animal" was spotted by a couple walking their dog in Queens Park, off Ferrybridge Road, in Pontefract at 18.00hrs. Their Staffordshire bull terrier "caught the scent of something" in the bushes, it went in to investigate, then came "running out at great speed, terrified."

The "husband" went in to have a look and exclaimed, "my god I haven't seen anything like it before," to the "wife." He told her that he saw a big Labrador-sized, grey cat with a long tail. At that the cat came out of the bushes mid-way so the wife could have a look and then turned and left. The couple said it was definitely not a normal cat. Their dog, who has a reputation for "teasing" local cats, was quite terrified of the animal. Now the lady will not go back to the area and has warned her 17-year-old son to stay away, and make a long detour if necessary if going around there. **(Source: BCIB).**

12th July: Seen by Gary Reed and two witnesses in a field off the Manchester Road in Slaithwaite, a cat with a jet black, glossy looking coat. The tail was long and the estimated length was the size of a fully-grown lion - 1½ sizes of an Alsatian dog. It was seen for approximately 5 seconds, and was on the other side of the road very close to main road - approximately 20m away.

Mr. Reed said: *"I was travelling on the 184 first bus service from Manchester to Huddersfield and the cat was eating grass. Let the cats roam - they have as much right as us to wander around. I have contacted BBC Radio Leeds about this sighting."* **(Source: BCIB).**

18th July: *"After reading your big cat diaries, I thought you might be interested in the sighting I had recently. It was approx 3am, and upon a visit to the loo, I happened to look out of the window - it was only a crack. Initially I saw* two cats, but they appeared to be massively different in size. I thought this was just due to the angle I was viewing them from, and the fact that I could only see properly out of the gap with one eye at a time. I opened the window fully, to see a small, domestic black cat cowering very low to the ground, moving slowly away from a very large, light brown/sandy coloured cat, sat near my garage.

It was sat upright on its haunches (is that the right way to put it?) And at first glance appeared to be a large dog, but as it got up and moved towards the domestic cat, it was unmistakably a large cat.

It was approx 2 ½ft high at the shoulder - it was next to a car, so easy to assess height, and had huge paws and a large fluffy tail (I can't honestly remember how long it was, I was concentrating on the cat itself). The main impression I got was that it was of a playful disposition, and quite a young cat, but I could be mistaken! It trotted across the car park in pursuit of the scuttling, ears back, terrified domestic cat, and disappeared from view behind the houses.

I have not heard any reports of missing cats on the estate, and no bodies have been found that I know of, so perhaps it was just playing! My colleagues reckon it was a Maine Coon, but it wasn't very fluffy apart from its tail, so I don't think that was it. Having seen the picture on your website of the cat running off with a rabbit in its mouth, I am of course comparing it to that, but the cat I saw was leaner - maybe it couldn't catch rabbits and was trying small cats instead? Does that count as cannibalism? Anyway, I will keep my eyes peeled, and let you know if I see it again. We live on Alanbrooke Barracks near Thirsk, and the car park backs on to open farmland between Topcliffe and Thirsk/Northallerton. There are large wooded areas, and lovely hedgerows to hide in, and plenty of deer and rabbits to feast on!" **(Source: *Ian Bond*).**

19th July: Sighting in the Drax area of Newshome at 01.30hrs when a motorist on the A63 near Newshome spotted a large black cat in the fringe of their headlights; it ran from left to right across the road. *"It happened so fast."* **(Source: *BCIB*).**

19th July: Carol Duke (55) and her husband Geoff from Sleights spotted what appeared to be a large black cat as it passed in front of them on the A169 just past Sleights bridge as they drove up the hill towards Whitby. **(Source: *Whitby Gazette*).**

19th July: A gentleman, who resides in Gilberdyke, was driving along the A63 (near Newshome not far from Howden) at 01.30hrs with his wife, on their way home from the Leeds / Bradford Airport. Suddenly a large dark / black animal ran from the roadside, at the Drax Village side and bounded across the road at phenomenal speed. The animal was caught in the fringe of the headlights and the gentleman caller is not positively saying that it was a big cat that he and his wife saw. He is amazed at the length of the tail of the animal, and states that it was as big as a Labrador dog. It was not until he saw an article in the *Goole Times* several days later that he thought maybe that it was, after all something out of he ordinary that they saw. **(Source: *BCIB*).**

22nd July: Sighting at Drax Village, Yorkshire at 06:30 by two witnesses from a distance of 150m. The cat was black with a long tail that started to curl at the end. Its height and length were equated to about the size of a Labrador. It looked like a black panther.

One of the witnesses said: *"I was driving a car. The cat jumped a hedgerow and entered a field."* **(Source: *BCIB*).**

22nd July: Seen at Drax Village. Nicole was walking her dog late at night, and as she headed towards the edge of the village near the last of the lights she saw just up ahead, a large animal shape coming out of the darkness near the football pitch. It stopped and stared at her, her dog then went berserk.

She then, as she got a little closer, realised she was looking at a big cat; brown / yellow in colour, as big as a Labrador dog, maybe just smaller then a Great Dane. It had a small head, and by the way it walked, and then just stopped, she knew by the movements that it was a cat.

Her dog didn't like it one bit. The animal just stopped and stared at her, as if daring her to go on. She stopped, turned and began walking slowly away. After a few paces the cat turned and slunk back into the darkness. No one in her family would believe her until a few days later when a neighbour said she had spotted the animal. When the BCIB article appeared in the local paper she felt vindicated and showed everybody! **(Source: *BCIB*).**

26th July: *"I have just read the article in the Selby Post about sightings of big cats in the Selby/Drax area. I live on Ousebank, which is close to the centre of Selby, but we have the river on the front and allotments and playing fields to the back. After the allotments, towards Drax, there is open countryside mainly, apart from farms and some cottages.*

On Monday 26th of July I was driving to pick my daughter up to take her to work. I left my house at about 6.10 am.

As I was driving up the road, which is between the allotments and the playing field, I noticed an animal in the road. It was just lying in the sun. As I neared, it stood up and jumped off the road into the bushes, which surround the playing field. As it stood up and before it went in, I only got a close look at the back of it, it was definitely not a domestic cat, and it was not a dog.

It was completely black and had a long straight tail, which was probably only about 6in from the ground. It appeared to hop like a cat. I didn't get a good look at its head, but when I got to my daughter's, I told her about it. I am pretty certain it was a cat, but the size of big dog. I am pretty sure we are on the opposite side of the river from Drax." **(Source: BCIB).**

30th July: Sighting by Adam McKale, and his colleague, in Drax Village. They are police officers, and were answering an emergency call, blue lights flashing as they raced along by the Camblesfourth roundabout towards Drax at 01.30hrs. As they came around a corner in the road, they saw, stood in a dip, a cat as big as a Labrador dog. The driver had to take evasive action to avoid hitting the animal.

Obviously being on an emergency call they could not stop to take a closer look at the animal, which they believed, was a panther.

Adam has kept our number and will ring should any more news come in, or indeed should he again see the animal. **(Source: BCIB).**

5th August: *"About 7pm last night, on the B4520 between Brecon and Little Chapel, we saw a very odd looking cat. As we were bombing along we didn't have a hope in hell of stopping to investigate further, but from the quick glance I got, it was near a settlement (can't remember which one - sorry). A big, bigger than a domestic, cat, in a red / brown colour. The hair looked quite long; I think it had a dark dorsal stripe with other stripes coming down the sides... sadly that's all I can remember as we were doing about 70mph at the time."* **(Source: Rachel Lacey BCIB).**

8th August: *"Telephone call by a chap who said he spotted a big cat near Snaith (near Drax) as he walked his dog along the river. He saw an animal up ahead of him and as he focused on it, he suddenly said out loud to himself (which surprised him) "that's a cat."*

It was jet-black with a 'long tail,' the cat looked at the man and dog for a few moments, weighing them up, then turned and 'just disappeared.' I asked the gentleman about this and all he would say was, 'that it just disappeared.'" **(Source: BCIB).**

10th August: *"Myself and my husband both saw a large cat at approx 7pm on Friday, 10th August.*

We saw it in the field opposite our house which is about 1 to 2 miles from Drax village, (Newland Village). It was large and black, and we do have some photos of it, although not very clear. My husband did cross the road for a closer look and is adamant it was not a dog, but was the size of a large dog."

Follow up:
"Emma and her husband were upstairs when they saw, in the field behind their house, a cat 'as big as a dog with a long curled up tail.' Emma's husband went outside and got 'fairly close,' then the cat 'crouched down, as if to stalk or pounce,' while looking at him. He then slowly backed away and came back into the house." **(Source: *BCIB*).**

10th August: Prints found in Sandholme, near Eastrington, East Yorkshire. **(Source: *BCIB*).**

11th August: Large cat spotted at Monkhill, Prince of Wales pits which are now closed.

"...The witness was walking through this area, which he shouldn't have been in, to be honest. Hardly ever used track, no one much goes down there, although building work started over the last week, as new houses are being built.

He came to a small clearing, and standing inside it was a big black panther standing next to a bush, just watching him. He stood dead in his tracks. The cat then casually turned and ambled off into the trees without a sound. He describes it as being as big as a fully-grown German shepherd, but longer in body, and as being jet-black with a "very long tail. The witness turned and went back the way he came, taking special notice of the dead half eaten pigeon he saw, and which he had taken no notice of on the way in." **(Source: BCIB).**

27th August: A big cat has been spotted in Ryedale. Sisters Lisa Farrow and Tina Wilson-Kallagher, both of Kirkbymoorside, likened the "big black creature" they saw in Ryedale to a panther. **(Source: *Malton and Pickering Mercury*).**

28th August: Anne Wilkinson Ravenscar was walking along the old railway line with her Jack Russell, Lady, at around 7.30pm when they came face to face with a large black panther-like cat, between Fylingthorpe and Ravenscar. **(Source: *Whitby Gazette*).**

28th August: Sighting in the Drax area at 11.00hrs. *"Ex-police officer John Anderson, from Goole, along with a friend are camping along the river Aire, at a little place called Templehurst where there is a campsite and a public house called the Sloop. John and a friend were sat beside the river when they saw a large black animal walking across the field. They thought it was too big to be a dog and too small to be a pony.*

They realised it was a cat. John describes the animal as being 'jet-black, walked with the definite gait, shoulder movements of a cat, but its hair seemed shaggy but short.' John observed that the animal looked 'under-nourished.'

The animal was not moving at any great speed, just 'making a B-line for a barn', which it went behind, and after that they never saw it again. John has just returned from holiday and knew nothing of the recent spate of sightings in nearby Drax. He mentioned the sighting to the landlord of the Sloop who replied, 'Oh aye, there is one around,' and never said another word about it.

Later this afternoon the two witnesses walked along the path they thought the cat had taken but found no obvious signs. At this very moment (19.45hrs 28th August), the two chaps are sat outside their caravan with binoculars, waiting and watching...." **(Source: BCIB).**

August: *"I read with great interest your article in the Selby Post, re: sightings of large cats. I have no doubt in my mind that what I saw on the A63 between Barlby and Howden was neither a dog or domestic cat, and it certainly did not have the gait of a deer. I drive that route very early in the morning and on this particular morning the weather was fine and the moon was bright.*

The animal crossed directly in front of my car left to right at no more than 50yds; it was close enough for me to react by braking. Its body was low to the road and its thick tail high and curved over its back. I had plenty of time to observe it as its walk was almost nonchalant as it made its way to the hedge that ran along the side of the road.

I hope this does not sound overly dramatic but the lasting memory of the sighting was the way in which the animal lowered its shoulder towards me for an instant and then was gone through the hedge." **(Source: *BCIB*)**.

August: *"After reading the article in* The Goole Times *today I thought I would pass on some information.*

Both myself and my partner have seen what we thought were big cats in this local area. My partner was travelling to work at around 5.00am on the A614 towards Kellingley, and he saw an animal pass quickly in front of his motorbike. He described it as a large black cat that looked like it had a monkey's tail that curled at the end, and it headed off across the fields to the side of the windmill.

Last year I also saw what I can only describe as a very large black cat with a long tail by the side of a small copse just off the A641 heading towards Rawcliffe Bridge. This was during the day, in bright sunshine and the cat quickly made off into the trees.

I am not sure whether we saw anything out of the ordinary, but they certainly were very large for domestic cats, as big as a large dog." **(Source: *BCIB*)**.

September: Two sisters from Kirbymorside spotted a jet-black' animal with a long, curling tail' as they drove along a back road near Oswaldkirk. **(Source: *Gazette & Herald*)**.

7th October: Robert Rogers from Poplar Farm in East Heslerton was checking crops in a nearby field just before dark on Sunday when he came across huge animal tracks on his field. **(Source: *Gazette & Herald*)**.

10th October: Sue Elsey from Leppington, between Malton and Sheriff Hutton, said she saw a black cat the size of a small Labrador as she drove along a quiet lane near her home. **(Source: *Gazette & Herald*)**.

28th October: *"We think we saw a panther at Higher Eastwood Todmorden. It was very dark in colour and very cat-like in the way it walked, with quite an athletic gait. It was roughly 26-30in in height. It was on an access road that runs through woods that hold roe deer. I have a flock of sheep above the woods, but have not been troubled by losses. It was on the edge of dark at the time. It scaled a wall of roughly 5ft with an immediate steep bank running upwards, from virtually a standing start.*

Out of interest have there been any other sightings in the area around Eastwood, Todmorden, Hebden Bridge?

BCIB representative Cheryl Hudson followed up the sightings and reports:

"I have been to see 'J' and family this afternoon at Todmorden. He and his wife were very, very shocked to see the black cat last Monday. It was around 4:30 p.m. less than ½ mile from their farmhouse. They have a ferret, 2 cats and a feral. They know the size of dogs, cats, deer, foxes, etc obviously. They judged the size (from less than 20ft away) to be between 32 – 35in high. The head was low in line with its shoulders and was very feline in its walk. They had, last week, scoffed at neighbours higher up the hill who told them they had seen a big black cat, thinking it was the feral that plays with their cats. They didn't notice the shape of the ears or the length of tail (though they said it had a tail, but didn't remember how long etc) because they were completely stunned. It was in a densely wooded part covering a huge area, and next to a railway line.

I looked up other sightings in the immediate area and there have been two in close proximity within the last two years. One a big black, and the other a lynx. The couple are so excited." **(Source: *BCIB*).**

19th December: Sighting at 6.45am along Whitehall Road (A58) between Drighlington / Birkenshaw of a black cat with long, thick, black tail, 18in long, 2ft in height approximately. The witness said: *"The length etc., was difficult to judge as it was vaulting a wall at the time 50yds away from me. I was driving down Whitehall Road when the cat appeared from the right hand side of the road, and loped across to the left side, and then leapt over, and cleared, a 4ft high wall. It then ran off across a field."* **(Source: *BCIB*).**

29th December: Report from Robert Mochrie of sighting at 15:45 by two witnesses in Tunstall. They are not sure, but it seemed to be black, but it was seen against sunset. It had pointed ears. Mr. Mochrie said: *"The tail couldn't be seen as the cat was watching us from a hill top. As for its height and length, we are again unsure as it sat on its haunches and the lower half was obscured. I would probably say 3-3.5ft. It looked like a lynx and we had two separate sightings.*

We initially drove past it and spotted it out of the corner of our eyes so we turned the car round at the end of the road and drove back, which was around three minutes lag. It was still there, so we parked up and watched it from roughly 20m away. It just sat watching us from the hilltop. The total duration was around 30 seconds."

Second sighting:

"As above, then during the second sighting we parked up on the side of the main road and sat and watched it. It knew we were there and crouched down, but we

could see the tops of its ears. It then lifted itself up onto its front legs so we could clearly see its head and neck. Then it pivoted round and disappeared over the brow of the hill.

The experience was awesome. It was a very strange feeling and definitely got the pulse flowing. It felt strange realising this cat knew we were there and was watching us. We are 100% confident that this was a big cat as we were sat in the car watching it not 20m away." **(Source: *BCIB*).**

30th December: *"I live in Drax village, Selby and regularly take out my dog for 6 mile bike rides along the nearby country lanes. I have heard of the sightings of large cats in the area and on 30th December came across the following evidence, close to the power station approaching Pear Tree Lane. The spot is where I have regularly seen roe deer in the past.*

I do not believe the carcass was due to poachers as the rump and the hind legs were intact and there were no sign of internals. Most of the damage appeared to be to the neck, body cavity, and cleaned off ribs. Unfortunately, I did not have my phone on the initial sighting when the carcass was very fresh. The photos were taken the day after when either the carcass had been revisited, or cleaned further by carrion. The kill being on vegetation, there was little evidence of prints." **(Source: *BCIB*).**

On several occasions, my cat has alerted me to 'something' being out there.

Scotland

Aberdeenshire
17 reports

3rd January: A mystery big cat is on the prowl in the northeast of Scotland, according to police. Special Constable Thomas Jennings, 24, came face to face, in the dark, with the 5ft-long black animal after he was called to clear cattle from the A944 Whitehouse to Alford road. **(Source: *The Scotsman*).**

23rd January: Witness Matthew Maines said: *"I live at Tillycairn, old Meldrum on farmland. I came out of my house at 09.00hrs to walk my dog with my young son, when something caught my eye running along the opposite side of the field close to the perimeter. I then stood and watched the creature disappear into the hedgerow in the corner, where it must have stayed for 10-20 seconds, presumably as there was traffic coming along the road that runs along the hedge.*

I then noticed it again running up the adjacent field on the opposite side of the road. Again, it stayed close to the stonewall which separates the two fields on the other side of the road. It then disappeared out of view once it reached the top of the field." **(Source: BCIB).**

16 March: Sighting at 5.00am in Rosehearty, on a clear night. It was dark in colour – but it was too dark to really get a good look. Their dog was howling at it through the window, and it moved very fast. It was a large unusual feral, and was seen for about 45 seconds at a distance of 200yds.

The witness said: *"It was the early hours of the morning, and we were woken by our dog howling at something in the garden. My husband went to the bedroom window to see what she was howling for, and saw the big cat in our garden. It was still quite dark and the cat moved very quickly."* **(Source: BCIB).**

17th March: Sighting at 6.30 am on the A944 Alford to Aberdeen Whitehouse. The ears were pointy and tufted but the witness did not notice a tail. It was brown and looked like a lynx, and was Labrador sized. It was seen for one minute whilst driving, at 100yds away.

The witness said: *"I have lived here all my life and have never seen what I saw yesterday – I still cannot believe it. I was driving to work and saw a lynx / bob* cat running at speed across the field - next to Lonenwell Garage – with powerful stocky movements." **(Source: *BCIB*).**

2nd April: Sighting at 5:00am, on the A87 Lochside, approximately 5 miles before Cluanie (heading towards Lochalsh). It was reddish brown, with a big bushy face with dark stripes, which appeared to be diagonal. The face shape resembled a racoon. It had pointed ears, and was seen for a few seconds - the closest it came was 2m away. It was beside a sheep and was smaller then the sheep, being the size of a large fox/or small collie-sized dog.

The witness said: *"I was a passenger in a car driving along A87. I saw eyes at side of road. As we got closer, I saw an animal as described above. It was at the Loch side and looked like it was eating a sheep (dead sheep and lots of blood!). Thought 'what on earth was that?' as I have never seen anything like it before and have travelled the Highlands all my life. The driver never saw it. None of the pictures I have seen whilst searching the internet look like what I saw - it did not have an obviously cat shaped face."* **(Source: *BCIB*).**

April: Suspected big cat claw marks found on a tree in woods. **(Source: *Graham Levy*).**

20th April: Sighting of a black cat at 11.00 pm on the A947, between Oldmeldrum and Turriff, possibly near Fyvie. It had a long tail, and was approximately 3ft tall, and 6ft long.

The witness said: *"I was driving friends home along the A947 to Turriff. Unfortunately, I am not familiar with the road so cannot give an exact location. I spotted a pair of eyes at the side of the road so slowed down in case the animal crossed. The animal walked across the road straight in front of my car so I saw it quite clearly in my headlights. It was a large black cat. There was no mistaking the shape of the head, legs, large paws or walking style. Unfortunately my passengers in the back of the car were not looking at the road so did not see it."* **(Source: *BCIB*).**

30th May: Two witnesses saw a large, unusual black feral at 7.00pm in the grounds of Forehill House, Blackhills, Peterhead. It had a long, sleek, tail arching towards the ground. It was seen for around 4-5 seconds at a distance of 100m. It was 60-70cm tall, and approximately 1m 20cms long.

One of the witnesses said: *"The animal moved in a very feline manner from left to right across our path, turned its head to look in our direction then went under a fence and into a field of oil seed rape. My son and I were in the car at the time of sighting."* **(Source: *BCIB*).**

14th June 2007: Sighting at 9:45pm in Bucksburn, Aberdeen, on the minor road to Kingswell's near the old people's home. It was black, with a long tail which was in the air with a curl at the end, and the hair on the tip of the tail appeared longer and a little bushier at the end. It was seen for 4 seconds at 50m, and was 5-6ft in length, and around 3-3.5ft high.

The witness said: *"I was driving up the road towards the corner, when I saw a large black cat jump on to the road and bound across to the other side on to the grass before jumping over the fence into long grass. The way it moved, and its size, meant it was definitely not a dog or domestic cat. This thing was huge and powerful - you could see in the way it moved."* **(Source: *BCIB*).**

21st June: Animal spotted at 16.30hrs on the North Side of the A96 Trunk Road, ½ mile west of Pitcaple by Keith Hay, at a distance of 30m. The sighting duration was 10 seconds. It was jet black, with no distinguishing features, and its tail was long in relation to the body, held aloft in typical domestic cat "swaggering along" pose. It was approaching the size of a Labrador, and was walking in a grass field with its body easily visible, so the witness suggests a height of 450-600mm.

Mr. Hay said: *"Whilst driving along the A96 heading west approximately ½ mile west of Pitcaple, I caught sight of the cat calmly "strolling" through the grass field on the north side of the road, heading directly towards the road. The cat was approximately 30m from the roadside. Unfortunately, as this is an exceptionally busy road, I was unable to stop or find a suitable turning point to turn around and investigate further. My sighting was fairly short, but what jumped to my attention was the fact that the cat seemed un-phased by the traffic on the road, it just seemed to walking along at its leisure.*

Other road users must also have seen it, its solid "blackness" stood out like a sore thumb against the bright green field. Until now I have always looked upon these sightings with some scepticism, but boy was I proved wrong today, the animal I saw was without a doubt a "big cat". My interest has been sparked and I'm now on a mission, I travel this road regularly and shall make every effort to obtain a further sighting. (PS: My wife now things I've finally lost it!)" **(Source: *BCIB*).**

9th July: *"Please could you have a look at these photos and offer an identification of what made them? They were found in a small patch of soft mud at the entrance to a field, alongside tracks made by a stoat. More tracks were found further up the farm lane leading away from the field. These were in a rural area just west of Aberdeen, at South Rothnick, which is near Netherley, between Aberdeen and Stonehaven. I noticed the shape and the lack of claw prints, which made me wonder what it was. I am an ecologist so am regularly looking for, and identifying, animal tracks, but couldn't work out what these are.*

Please could you respond as soon as you can as I am quite intrigued and need to know whether I should hide this from my colleagues if they are dog!

Similar prints were found last year near during badger surveys around Stranog Hill.

I spoke with my colleague, who found more tracks leading up a farm drive and he thinks there was about an 8-inch gap between each print. This wasn't measured so is an estimate." **(Source: *Chris Kerfoot*).**

14th July: Sighting by two witnesses at approximately 11.15pm at Cruden Bay, very near the village just over the main bridge going into Cruden Bay from Peterhead Road. The animal was black, with a long, strong-looking tail with a slight rise at the end. Could not say about height and length.

The witness said: *"We were under a meter away, but only saw the back of the beast before it disappeared into wooded area. I was driving home from Aberdeen, and we turned the corner out of Cruden Bay, put on the full beam of the car lights and the animal had been crossing the road. It just continued into the thick undergrowth with a little leap."* **(Source: *BCIB*).**

9th August: Sighting by Mr. Albert Handsley at 5.35am near Potterton. It was light brown with a brown face, and spotted pointed ears. Its tail was 3ft long and curled at the tip. The animal was approximately 3ft long - 2.5ft high.

Mr. Handsley said: *"I was walking the dog in woods when I spotted the cat running. It turned, looked for an instant than ran off. The cat was silent; these cats should be left alone to do what they do in the wild."* **(Source: *BCIB*).**

25th August: A large black cat seen on the outskirts of Keith, possible cubs - ongoing investigation with cameras. **(Source: *Di Francis*).**

1st September: Large black cat spotted near Keith. **(Source: *Di Francis*).**

8th September: More sightings of a black cat near Keith.
(Source: *Di Francis*).

17th November: Sighting to the rear of Uppermill Farm near an isolated cottage, surrounded by fields, 1 mile along Tillycairn road (1½ miles from Tarves village). There were no more witnesses, except a very frightened pet cat of the black animal, which had no obvious markings. It had a long slimmish rounded tail, and was smooth haired, with no feathering, and was about ½ the length of the body. It was held horizontally from the rear with slightly curved up end. The animal was taller than a large black Labrador, but longer in the body, and had a short haired coat very similar to this breed. It had long sinewy legs which were not thickset at the shoulder, and a lithe and muscled body (grey hound-ish). Its head was viewed from side, and seemed quite small, and the neck was short and muscled. The witness did not think this was thickset. It was a large unusual feral.

The witness said: *"I was slowly driving along cottage track with full beam, on a very dark night. My pet cat ran up to the car, and looked alarmed and kept looking behind him. As I stopped to pick him up, I saw, in the car headlights, a large black animal appearing from my garage and loping away down the side of the track to the right side of the garage, and into a field behind it. It did not stop, or turn to look, so my view of it was from the right side and rear of the animal.*

I was inside my car, and the engine was still running, so I heard nothing. My pet cat, a large outdoor tabby, jumped into the car beside me and was lackeyed and frightened and kept surveying the garage area when I let him out - he has since been very unwilling to go outdoors unless I am with him and constantly surveys the area from the back door steps.

He has been avoiding the garage, even though he is normally an outdoor hunting cat and has his bed in there under the eaves, and has taken to hiding under my camper van parked not far from the garage. This state of alarm has been going on for the past week since he appeared out of the darkness one evening, one side of his body soaking wet and covered in mud and with a round gash on his head - now I have seen this animal it may be a possibility that he was swiped by it. His appearance and fear suggested something traumatic had happened to him at the time, but I was not aware of any presence of the black animal.

Tonight my cat refused to go out, and when I returned from an outing at 8.00 pm, I saw a small black cat lying under my camper van - it appears to be dead - but as my young son was with me, and it is very dark outside, I have not gone out to pick it up.

The day after I saw this black animal, I let my collie dog out and watched to see what he would do. He went straight toward the garage and followed the scent from there along the track, and when I caught up with him he was following the scent along the hedge line of the field beside the drainage ditch which was the direction the animal had gone.

Tonight I saw what may have been the reflection from an amber eye shining from the stack of bales behind the outbuildings of the farm as I drove past to my cottage - just before I saw the dead cat under my camper van.

I feel alarmed and nervous at the thought of its presence, and the effects it seems to have had/is having on my pet cat and dog, as they are still very wary right now. My son does not know, and I would not want him to just now as he is already afraid of the dark and would be very afraid to continue living here knowing of the presence of such an animal. I would appreciate if there is someone who could come along and have a look around for traces of the animal and to check out that poor wee dead cat to see why it has died there.

If it is to do with this animal, and it is still out there near this cottage, or around the farm and outbuildings, then I would hope it could be traced, caught and removed." **(Source: *BCIB*).**

Angus & Forsarshire
9 reports

8th February: Mr Peter Laing spotted a big cat when it broke cover from the side of the road, and ran right across the path of his car. He said: *"I was between Farnell and the straight which leads down to the Brechin bridge when this thing came right across the road and into the woods."* **(Source: *The Courier*)**

15th March: A family from Pitkennedy, near Forfar, are claiming to have evidence that there is a big cat in the area. Two large paw prints were left in the back garden of the Lyon household at Turin Cottages, and daughter Shelly saw a cat-like animal the size of a dog. **(Source: *BCIB*).**

7th April: A two minute sighting of a large black cat at 10.00pm at the Millennium Park, Carnoustie. It was seen at a distance of 20m.

The witness said: *"I was out walking my dog, and I was looking up at the stars, walking slowly. The dog was a bit behind chasing a rabbit. I stopped just before the crossroads of the 'dog walk' and shone my torch to see if I could see any owls. At that moment, I saw something shining, so I turned my torch back to the spot and there was a pair of large green eyes looking straight at. I then turned my torch to the left of me and there was another pair exactly the same. I was scared, and I hit my torch with my stick, and shouted to my dog who came up running. I started to walk backwards, and I shone my torch again, but the eyes had gone - it was not or a fox as they have different eyes."* **(Source: *BCIB*).**

April: Sighting in the grounds of Stracarthro Hospital, off the A90, the nearest town being Brechin.

The witness said: *"The weather was good. It was early morning, in the spring, when my wife Janette was going to work on an early shift at the A & E department in the hospital. The animal had rounded ears, and was sighted for approximately 1 minute at a distance of approximately 150m.*

She was driving in the hospital grounds going to the car park, when the cat was sighted walking through the woods adjacent to the hospital grounds" **(Source: *BCIB*).**

20th April: Sighting at 7.10am of two dark coloured cats, on the outskirts of Arbroath, just off the Arbroath-Dundee dual carriageway. They had long tails, and were seen for about 10 minutess at a distance of 50yds, before running diagonally out of sight over three fields. They were roughly similar in size to a

Labrador dog.

The witness said: *"I was driving along the road, when I saw a dark object bounding through short grass in a field to my left. I looked again and there was another around 20yds behind. I then pulled over, and watched as they ran diagonally across three fields. I was also looking to make sure they were not dogs, or if anyone else was around. As they got to the edge of the third field, two deer sprung up in the air and started to run, with the cats in chase.*

I went past a family friend who lives opposite from where I saw the cats, as she had one previously in her woods. This week I went out to see her son who does a lot of game shooting, and he has seen them five times in the last three weeks. He does not have a photo, but is now carrying a camera at all times. He has, however, identified that they are a more red/brown colour.

In the woods that they own they have seen a considerable drop in rabbits, and also a few weeks back he put a deer carcass that he had been hanging for maggots, only to discover the whole deer had disappeared." **(Source: BCIB).**

Early August: A couple spotted a large cat with "*a glossy coat, yellow eyes and sharp pointed ears,*" behind Dundee's Ninewells Hospital. **(Source: *Dundee Courier*).**

October: *"In the early hours of the morning, I recall hearing outside; what sounded like a panther or other big cat roar a bit, as though it was challenging another animal. At the time, I didn't think much of it, but it became more apparent later this month when just remembering, that the sound I heard was definitely that of a puma or jaguar.*

It was north of Monifieth near A92." **(Source: BCIB).**

2nd December: Sighting by two witnesses at 3.30pm, on the Longforgan to Fowlis road 5 miles from Dundee. It was a black cat - very muscular - and bigger than a large dog prowling through the field. It had a long tail. Rachel Flynn said: *"The cat was on the top of hill directly in front of us as we approached in our car. It was prowling, then disappeared in to the bushes. We turned the car around and got out with a camera phone, but couldn't see it again."* The Dundee Courier reported the sighting on 3rd December. **(Source: BCIB).**

3rd December: A female motorist travelling to Fowlis spotted a big cat, bigger than an Alsatian. **(Source: *Dundee Courier*).**

Late December: Golfers at the Camperdown Golf Course in Dundee found large prints in the bunkers. They believe them to be cat prints. But after viewing the photographs, it is BCIB's view that these are dog. **(Source: *Sandy Smith*).**

Argyll & Bute
6 reports

4th January: *"We get occasional reports of a brown puma-type cat in the area. In fact, one of the most compelling eyewitnesses I have spoken to describe a brown puma-type animal at Bellochantuy, Argyll.*

Well I can now confirm its existence with my own eyes. I've just been for a drive around the usual hotspots in the area, and was driving along the coast road at the tip of Kintyre, at about 4:50 pm when, in my headlights, I saw a large light brown / fawn coloured animal jump a stone wall (with barbed wire on a fence attached to the other side of it, and no, there were no hairs to be seen on the wire), at the side of the road to my left. It cleared the wall with one bound. Although I didn't get a good look at its head, the thing that struck me was the muscular shoulders and the tail. The tail was thick and long, (and although it was slightly curved upwards, I wouldn't say it was the typical 'S' shape) and as it jumped the wall I could see the tail hang down at least half the height of the wall.

I immediately stopped the car, grabbed my torch and searched over the wall. Unfortunately, it had disappeared out of sight. You could just make out the tree line on the other side of the field in the fading light, and I thought I could see a movement through the trees, but it could have been a trick of my eyes, I couldn't be sure. There were sheep in the field, but they were all huddled at the opposite side of the field to my right.

I measured the height of the wall at 4½ft, with the verge at 2ft. The animal cleared both with one leap. I don't know where the cat had come from, because the other side of the road is a very narrow strip of beach and rocks leading to the sea. However, seals are known to congregate on these rocks in large numbers.

I did check the beach for prints with my torch, but all I could see were two tracks, one of a dog, the other of its owner, I presume. Unfortunately, there is no point returning to the spot tomorrow in the light, because the sea would have washed any tracks away.

Using the known height of the wall, I would estimate the animal was at least 4ft in length (possibly more) with a tail of at least 2ft. I've got a large black Labrador, and I've just measured her tail, which is just over 12in. This would rule out the animal being a dog in my opinion. I'm a little annoyed that I didn't get a good look at the head, but the arched muscular shoulders and long thick tail, could be nothing other than a cat of some description in my opinion.

I've included a drawing of what I saw, the location is less than a mile from

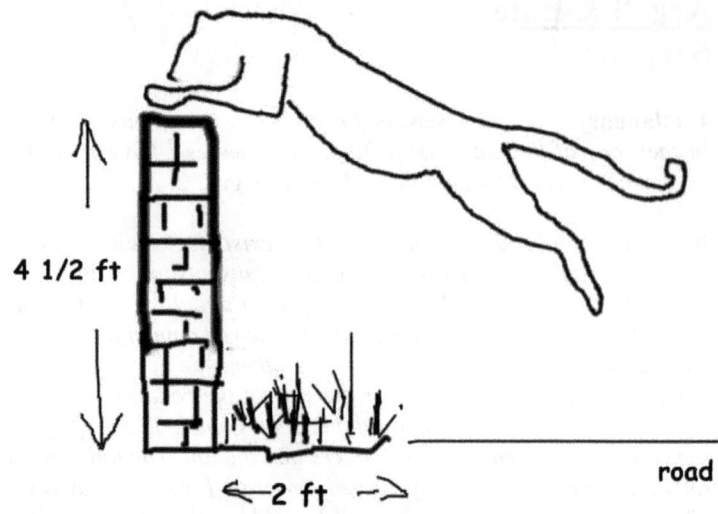

Dunaverty Golf Course, where a "jaguar" has been seen on a couple of occasions. The location where the puma was seen last year is about 18 miles away as the crow flies. This sighting today, however, is the first time I've heard of a brown cat being seen in this part of Kintyre." **(Source: *Shaun Stevens BCIB*).**

6th February: Sighting by John Gallagaher at approximately 8:15am at Machrihanish, Argyll. It was 3°C and just after dawn, with clear skies. He saw a large jet-black cat (smaller than a Labrador) on the edge of the beach, sniffing at a white carrier bag. The animal was approximately 3ft in length, and 18in in height. It had a curved tail, which was approximately ¾ the length of the body. The witness' dog spooked the cat, and chased it into reed beds.

Follow up by Shaun Stevens:

"John came round to my house just before 9am to tell me of the sighting. I was on site by 9:30am. I found the carrier bag he mentioned, which contained the remnants of a Chinese meal, and bits were scattered about. We only have about 4 days of frost a year, and last night was one of those, so the ground was solid. However, about 20yds away on the beach I could see a number of tracks. Amongst the dog, sheep, goat and human tracks, there appeared to be a set of cat tracks. Photographs were taken of those tracks. The paw prints were about 2in square, and about 24-26in apart. (For comparison, on arriving home, I got my Labrador to walk through a puddle, and then onto a dry patch of concrete, and her prints measured about 22-24in apart.

This is another sighting, not of a leopard or known species, but one of this group of large ferals/hybrids/unknown species, that we are getting regular reports of in the area. My first impressions of the tracks were that they were quite small, although you must take into account the ground, including the beach, was fairly frozen, so it may be that the impressions made in the compact sand, may just be the tips of the cats pads. The interesting thing to me was the distance apart. I did an experiment with my black Labrador at home, and the distance between her paw prints while walking, were slightly less than these cat tracks were measured at. I find this very intriguing. My Labrador measures about 3ft in length, similar to the witness' claim of the size of the cat, and the length of the tracks of the two animals were very similar. This seems to back up the witness' estimate of size. Although the pad marks indicate domestic, the spacing between prints suggests a much larger animal with a much larger stride. What this animal actually is, I'm not sure...but it's certainly a lot bigger than a domestic...and I have no doubt it's out there." (Source: *Shaun Stevens BCIB*).

Photographs of scat and prints.

White card is 3" square. Found lodged in a narrow low section below a jumble of boulders, where large black feline has been regularly seen.

8th February: Sighting in Oban at 4am. of a dark brownish animal with blotches of orange, pointed ears and a very long and bulky tail. The witness described it as a large unusual feral of about 2ft tall and of over 1m in length.

The witness said: *"I was driving home from work, when I saw a large figure with a rabbit in its mouth so I stopped the car, got out, and approached the animal. As I got closer, it turned towards me, and then I realised it was a cat. I tried to call it over, then it ran off and I couldn't see it any more so I got in the car and drove off. The cat killed a rabbit and was very aware of me and obviously felt exposed so it ran off."* **(Source: *BCIB*).**

25th March: - Sighting by three witnesses at Tobermory. The animal was black, with a fairly thick tail, about 1-2 ft long. The cat was seen for about five minutes from a distance of 150m. It was 3ft long and 2ft high.

David Hancock said: *"We were driving while on holiday and the cat was first seen walking along the edge of a field into some woodland by a passenger. We took some pictures, and a video, which has geese in the foreground to give some perspective."*

Shaun Stevens reports:

"At the moment we don't have much to scale the cat against, apart from the telegraph pole. However, today while driving, I kept noticing the same yellow warning signs on the telegraph poles as I drove round the country lanes. They were on average about 7-8ft above the ground. I measured the signs, and they are 30cm high and 20cm wide.

I came back and checked the photo again, and by scaling up the yellow sign, it is 2.4m high off the ground - about 8ft - which ties in to what I've seen today. Now, looking at the cat, if it was right next to the pole it would be a least 12in high, which is about the same as a domestic. But it is quite a distance away behind the pole, so it is well above the normal size of a domestic."

Shaun also received the following stills, and we are now waiting for the actual footage. **(Source: *BCIB*).**

24th April: Shaun Stevens has had trigger cameras out in the area of Southend for several months. Among the deer and fox pictures, on the 24th of April, he caught the picture of a cat. Not a leopard or any other such foreign exotic as that, but a snubbed nosed black 'wildcat' so oft reported in this area. Could this be a juvenile of the British indigenous species that Di Francis advocates? She predicted that we would be seeing the juveniles around this time of year. Shaun Stevens' hard work may have finally paid off when he developed the film on the 8th of March.

Note the small, pinned-back ears, and long sloping forehead, and compare it to the two lower pictures.

The first is the animal found outside of the Bovington Tank Museum in Dorset by Jonathan McGowan - the second is an illustration by Di Francis created before the two pictures below were taken.

After the film was developed Shaun headed straight back to the spot to make measurements: He reports: *"All the bait has now gone, and a dead mouse, (recently killed, with bite mark to the neck) left in its place.*

Unfortunately, I'd left the trigger camera on standby when I last changed the film, and no new pictures were taken. Anyway, I'm going to re-bait the spot, and put an additional camera there as well. See what happens. If I can continue to get pictures of the animal, it might be worth setting up a small cage in the future. But I'll tread carefully on this one, as there are active deer hunters in the forest, and I don't want the animal killed if I can help it.

After carefully measuring the spot, it appears the cat is about 39in from nose to tail. Not a panther of course, but still a very large animal - much bigger than a normal domestic.

It's obviously the animal that everyone has been seeing in the area, although it is slightly smaller than they claimed it was, but that is to be expected."

The picture below illustrates Shaun's measurements.

After Shaun made some larger prints, he noticed that the shape of the head may have been distorted by the shadow, and after careful examination noticed that the shape of the head was as below. **(Source: *Shaun Stevens BCIB*).**

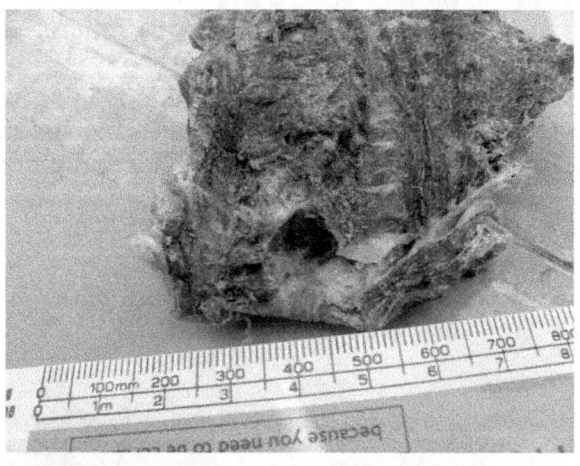

28th May: Sighting by two witnesses on Monday evening at Torosay Castle estate, Craignure, on the Isle of Mull. The colour of the animal was black/dark, and it was estimated as being as large as a collie dog.

On of the witnesses said: *"It was not actually myself who had the sighting but two of my neighbours. I was directed to your website from my online forum where I was posting about what happened last night.*

My neighbours sighted a 'large black cat' at the far end of the field opposite their home. They at first thought it was my black domestic cat, but when the husband went outside with the binoculars, he said it was a large non-domestic cat as big as the farm's male collie dog.

I believe that they watched it for sometime but do not have any more details. We both live on the estate of Torosay Castle, Isle of Mull. It is also a working farm estate and has both Highland cattle and sheep. In recent weeks I have been hearing strange cries during the night, which I do not recognise as being any animal I have known of or to be a bird such as an owl. There are no foxes or badgers on the island. I can only describe the cries as a cross between a howl and a screech, quite drawn out...really hard to explain.

Although these sounds have unnerved me, I did not really think much about them until today when my neighbours told me what they had seen. Both my neighbours are very 'ordinary' and down to earth people, not prone to looking for anything unusual.

I was surprised to read on your website that you are planning a vigil here on the Island soon and that there has been a sighting as recently as March! Please could you give me any details of big cat sightings on the Island? I would be very interested to hear about them and to help in any way I can." **(Source: *BCIB*).**

Ayrshire
9 reports

16th January: Sighing at 23.30hrs by security officer John Golder while on patrol at the Manse Estate in Galston, when he spotted a large animal he thought resembled a cat. He said: *"I shone my torch on to the playing field and immediately lit up a set of eyes, which then stopped and turned to look directly at my torch beam. Immediately the animal began to bound towards me at great speed. I became spooked, turned round and ran into the porta-cabin, blocking the door with a metal bar."*

John immediately rang Mark Fraser of BCIB, knowing Mark personally and of his interest.

Mark said: *"John said the animal "bounded" which reminded him of a cat, in fact he said he did not know any other animal that would move like this. When I saw him at 00.15hrs that same evening, he was still shaking. He obviously saw something that alarmed him, but we are not sure what. John remarked that the animal's eyes were yellowy / blue and bright - almost like a star - and seemed to be about 2½ - 3ft off the ground.*

John stayed in the cabin for twenty minutes before venturing outside again, but before he did he turned all his lights off and made sure he had a good look outside before he opened the door.

This area of Galston is actually rich in sightings stretching back over several decades. In fact, it is not the first time John had seen this animal. Several months earlier, he saw the same thing, but this time it just walked across the field." **(Source: *BCIB*).**

7th February: Chic and Cat Blades, owners of PB Security Contracts based in Kilmarnock, were driving along by Dundonald at around 23.00hrs. They saw a large black cat run across the road in front of them. Chic said: *"It was as large as a dog, completely black with a long tail."* **(Source: *BCIB*).**

23rd February: Sighting in fields in front of the cemetery, and beside Newmilns ski slopes.

The witness said: *"It may be trivial, but I thought I would drop you an email and let you know what we have noticed over the last couple of weeks. We have a large golden retriever dog that for the last couple of weeks hasn't been very happy, and often spends ages looking across at the fields opposite our house. What she sees or hears is beyond us.*

Well yesterday, our eldest son took her for walk into the field as there were no animals in there. She picked up the scent of something and was searching very quickly across the field, and eventually kept going till she reached the top of the field. She then did something she hasn't done before, but went through the hedge into the adjoining field. Where at first my son thought she had found another dog. On closer inspection he realised it was a dead sheep, which had basically been ripped open.

It is in the field just below the graveyard wall and two fields along from the Newmilns ski slope. It was only when we noticed from our window that something was wrong that we rang our son on his mobile and asked him what the problem was. That was when he told us about the sheep, and we could see the sheep from our window. It's not as if it could have been done by a stray dog, as there are none around here, and also all the local dog owners accompany their dogs on the walks.

This sheep has definitely been ripped open and some of its innards are hanging out.

This may be of no help at all but I thought it was worth dropping you an email and letting you know." **(Source: *BCIB*).**

29th February: Sighting by Thomas Anderson at 2-3 pm in Patna, New Bridge. It was black all over, with pointed ears, and was observed for 10-15 seconds from 10-15ft away. The cat was about the size of a large Alsatian.

Mr. Anderson said: *"I was taking a short-cut across a field, on my way to Doonbank, when, on the way across the field, I saw something black that caught my eye, and I knew it was too big to be a cat or a dog. It turned and saw me and it ran into long grass. I ran to try and see if I could get a closer look, but it had disappeared almost immediately."* **(Source: *Mark Fraser*).**

22nd March: - (two sightings) One at 7.30pm ish on the A78 to Mauchline (turn right at Crosshands B744 and then the first turning on right into single-track road). It was at the first house on left - Lakeside Cottage sits in front of the railway line on the back road through Carnell Estates. Witness was unsure, but just saw eyes that were dark coloured.

The witness said: *"I had observed, on Thursday, an eaten bird in the front garden, but noticed that the paw marks around it were larger than my dog's. Then stepson spotted it in backfield on Friday night - unsure - just saw eyes/dark coloured, but it was bigger than my collie dog.*

On Thursday at 4.30am, I was awoken by my dog, and thought it just needed the toilet. So I let the dog out the front, and her tail started to wag and I let the dog back in. I thought that was strange, following the morning when I noticed the half-eaten bird in the front garden, but when I went to look I noticed large paw prints and skid marks as if the animal had corned bird. My dog sniffed but did not go near the bird. I noticed faeces that are not my dog. I thought it may have been large cat. On Friday night my stepson went out back to let the dogs out and noticed the eyes of a large cat as he was putting on the spotlight. He stayed in the porch and the cat looked right at him, and then ran into the field out of view." **(Source: BCIB).**

17th April: *"Hi, my name is Ashley Jay, and I live in Kilwinning and was visiting my boyfriend in Stewarton, Ayrshire last night (Tuesday, 17th). I was travelling on Dalry road on my way home about 23.20pm, and thought I saw a black Labrador on the road ahead. I slowed down, but what I saw made me go rigid and virtually stop the car. To my surprise, this animal slinked of the road into a row of trees. It was the shape of a panther and moved like a cat, with a long tail. I stopped the car but it had gone. I feel daft as I am very sceptical of these sightings, but I thought you might be interested."* **(Source: BCIB).**

3rd May: Lampers (fox control) were just outside of the Eglington Country Park, Kilwinning at 01.00hrs, when they saw a jet-black cat through the night scope (?) at a distance (normal vision) of a 100yds. On of them said: *"It was quite a lot bigger than a fox, low to the ground with a long tail."*

Colleagues of the witness believe that there is a "tawny brown puma" in the area. Also the witness once found "cat tracks" going over a freshly seeded field. **(Source: BCIB).**

Banffshire
8 reports

1st January: A large black cat spotted by a man staying at the Springs Hotel, Banff. **(Source: *Di Francis*).**

30th January: Banff man, Stanley Bruce, claims that as he was walking his dogs beside the Red Well during the morning when he spoke to a man who said that the cat was spotted by a woman earlier that day. **(Source: *Banffshire Herald*).**

29th January: Richard Empson saw a large mysterious cat near his Banff home. He said: *"I'd been outside the house, walking round the front. It came down the side of the neighbour's house, wandered across the lawn and into the undergrowth round the side. I thought, 'am I seeing right?' It was big, and definitely not a dog; about 3ft tall and 4-6ft long with a smallish head – not a big* *head for the size of it."* **(Source:** *Banffshire Herald*).

28th January: Darren Thomas (19), of Seafield Street, Banff, was driving to Whitehills, taking two of his mates home after a night at the Seafield nightclub in the early hours of the morning. The cat crossed the road heading west between the Whitehills cemetery and the village of Whitehills. Darren said: *"At first I thought it was a dog, but it was too big for a dog, it was about 4ft long plus a long curvy tail and about 2½ - 3ft high. I never saw it for that long, it was only for about two or three seconds. It ran across the road. I never really saw it until it was on the left side of the road, right in front of me. It was pretty big, bigger than a normal cat. I got a bit of a shock."* **(Source:** *Banffshire Herald*).

8th January: Sighting at 15.30pm on the Old Banff Road near Portsoy. The animal was black in colour, and able to clear a 3ft fence at the side of the road without touching it. It had rounded ears, and had a long tail - about half the length of its own body - with a pom-pom type end to it. It was seen for 30 seconds to 1 minute from about 10 – 11ft away. It was about 3ft high and about the length of a large dog. The witness said: *"I was driving to my home when I saw the animal come out of the field into the road in front of me, I stopped and the cat looked at me then jumped over the fence into the woods."* **(Source:** *BCIB*).

February: A large black spotted by the Glenlivet distillery. The witness believes it was a lynx. **(Source:** *BCIB*).

25th March: *The Banffshire Herald* **reported:** *THE Beast of Banff, the mysterious large cat-like creature spotted prowling in the area, has been captured on video. Now you can see the footage exclusively on the Banffshire Journal's own website. The footage comes courtesy of a Banffie reader from the Cornhill area, who does not want to disclose his identity or the exact location of the sighting.* **(Source:** *Banffshire Herald*).

Early September: Keith from Banffshire saw a large black cat running across the road, at the Mill of Towie. Behind the [...] shows playing field. It jumped over a hedge, and had a three-foot tail. **(Source: *Di Francis*).**

Berwickshire
1 report

3rd February: it was during the late afternoon but had been blue-sky day - bright sunny conditions. The animal was completely black and looked like a leopard. It had rounded ears, and the tail was about 3ft long - thick in nature as well as long – which was curled at the very end almost into a hook. As the leopard moved down the bank, the end of the tail was nearly vertical in the air with a hook towards the head. Its height was about 3ft, and roughly 4ft in body length plus the length of the tail. Conservative estimate was 1½ times the body length of Labrador.

The witness said: *"I was walking very quietly with my Labrador on a lead down to the river through some woodland to look for otters. There are eight wild sheep that run pretty wild in the wood. They were just beyond the sighting, and I suspect the black leopard was interested in them. The leopard ran at a graceful sleek pace down the bank down a sheep track before disappearing over the brow. It was surprised to see me. The ground was hard so no footprints other than those left by the sheep from previous wet weather. It was extremely graceful and silent. The sheep beyond appeared to have held their ground before moving down the bank and around past me to safety."* **(Source: *BCIB*).**

Caithness
2 reports

8th May: A lamb was found by farmer with its whole head ripped off (Granton Mains). He believes it was done by an animal that he is not familiar with. Two more lambs were attacked. The lamb with the head missing had a cut that was clean, nearly "surgical." **(Source: *BCIB*).**

9th July: Sighting by two witnesses at Tofts of Tain. The animal was black, with no markings. It was observed for 5 seconds at a distance of 50m. It was 800cm long excluding the tail, and 400cm high.

Dave Gunn said: *"I was driving along from Tofts of Tain towards the Wick Castletown road with my wife. We both registered movement in a field to our right. A large black cat came into view moving very fast. It wasn't running away from us - it seemed to be chasing something. It didn't move like a domestic cat or a dog.*

We have no idea what it was. It certainly wasn't a dog and it was much bigger than a domestic cat. It had a sleek elongated body and moved very fast. The area where we saw the cat is pretty remote and the place is teeming with rabbits, so I guess that an animal like that could survive undetected for a long time. I don't see the need to do anything about them." **(Source: *BCIB*).**

Dumfries & Galloway
5 reports

21st January: A large cat "sort of black" in colour was seen at 17.00hrs near Beltonhill Farm near Dumfries. The witness thought it resembled a lynx - paw prints were found but not sure what they are. **(Source: *BCIB*).**

28th January: Sighting near Castle Douglas, when the witness spotted a dark grey, large cat, which he though might have been a lynx, but it had a tail and was the size of a Labrador, also described as a "scruffy thing" clawless paw prints also found. **(Source: *BCIB*).**

7th April: Sighting by Shona Cameron at 18.30, near Ae. She said: *"It was a warm sunny summer's evening. I'm positive this cat was a lynx. I looked up photographs when I got home. It had pointed ears, but I didn't see a tail. I observed if for about five minutes. It was 3ft from the car, right in front of us*

We were driving through Ae village, just after the village going towards Closeburn/Thornhill direction, when the cat came out of the forest to cross the road. He/she stopped in the middle of the road and froze for about 4 minutes looking at us then jumped over a dyke." **(Source: *BCIB*).**

June: Sighting by three witnesses at 05.00hrs in the Shore area of Annan. The animal was ash grey base colour with black markings, turning to a light sand colour on its under belly. It had pointed ears and a striped, mixed coloured, tail. The length of this was long for its size. It was spotted for two minutes at a distance of 100m. One of the witnesses, John Irving, said: *"I was not able to determine its size due to the speed at which it travelled. I was on a morning walk before a trip to a campsite with friends. That beast was scary - it walked down a public footpath before clearing a 6ft hedge."* **(Source: *BCIB*).**

1st July: Sighting by Christine Smith at 1.30 am at the junction of New Galloway Road B7079 and minor road by the cemetery of the A75 at Newton Stewart (Grid Ref NX 421 653 approximately). The cat had no markings, and appeared to be a dark tan, almost mahogany, colour all over. It had a thin, 'cat like' tail reaching nearly to ground and curved at tip (see sketch on the photograph). It was short and was observed for approximately 1-2mins, at a distance of approximately 20ft, when first seen. Its height was about 3ft - see sketch. The witness was not sure of its length.

Ms Smith said: *"I was driving along NW towards Newton Stewart on B7079 late at night in heavy rain at 30-40mph. The cat appeared on the opposite side of the road and crossed in front of the car approximately 20ft away (far enough that I did not need to brake). The cat then moved in front of bollard (see photo) and it was then that its size became apparent. The cat's movements were graceful and fluid. Despite the heavy rain, it was not noticeably wet. Please note that a puma is the closest ID from a quick look at pictures, as I did not notice head details."* **(Source: BCIB).**

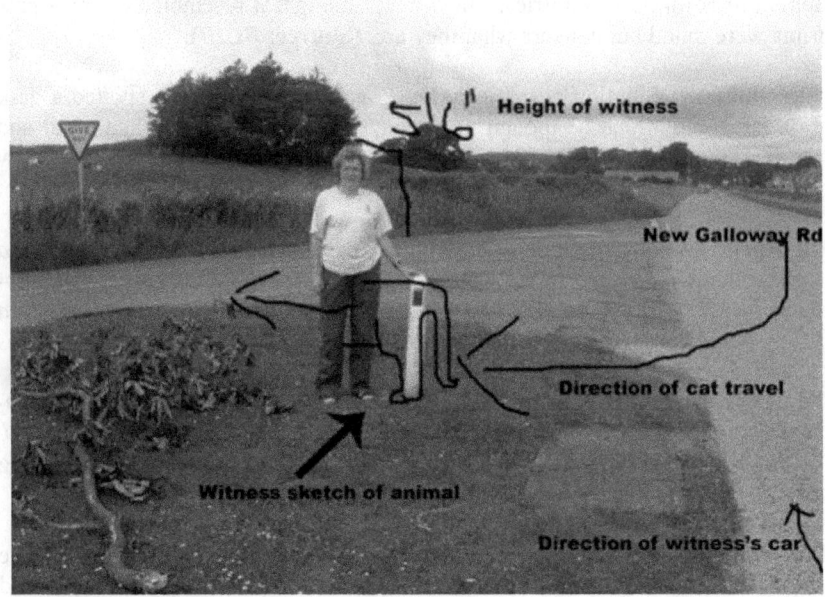

Dunbartonshire
2 reports

17th June: Sighting on the south bank of River Kelvin in Kirkintilloch. It was fine and cloudy.

The witness said: *"It was black, as far as I could tell. The tail came from its back and dropped nearly to ground level and then swooped back up. It was seen for 2-3 minutes from approximately 20ft, and was 1m in length and about 400-600mm tall, approximately. We were fishing for brown trout on a river. Obviously, we were being quiet and during the evening we had seen kingfishers feeding, foxes coming to the riverbank for water, and ducks and ducklings. Then, at 22:15, it was getting dusky and we saw an animal moving along the top of a wall in a builder's yard. It was definitely a big cat. I have a large male Maine*

Coon (domestic cat) as a pet, and the thing we saw was at least 3 times the size of him. The cat was on the opposite bank of the river and it moved the full length of the yard on the wall then returned along the same path. Then it came back for the third time. I tried to take a picture, but it didn't turn out. We then packed up at 23:00 and headed to the car. After one final look at this animal we waited near the yard and saw what looked like the same type of cat, but a little smaller. This was on our side of the bank, but at the roadside. When it saw us it disappeared into the bushes.

The cat made a sort of a noise like the type of half meow/half growl. A bit like the noise a domestic tom cat makes when another cat is in its territory." **(Source: *BCIB*).**

Early August: A Kirkintilloch man, who asked not to be named, was walking with his dad and two dogs when they spotted a mysterious cat down by the River Kelvin, at the back of Waverley Park. He said: *"Our Labrador went running into the long grass and that's when we saw the cat."* **(Source: *Kirkintilloch Herald*).**

East Lothian
1 report

28th - 27th August: Two sightings at 06.00hrs at Dunbar, in the Lochend woods, adjacent to Lochend road. It was very light gold, almost white, and had a long tail which was not bushy. The witness said: *"Initially, I thought what I saw was a Labrador dog (or similar) but when I got closer I realised it was a cat. Two sightings took place:*

Firstly: A very fleeting view of a large animal running across road and disappearing into undergrowth. I saw this from about 20yds, but the light was not good.

The second sighting was clearer and lasted about 20 seconds. It was daylight, and it was seen from about 20yds again. This second sighting was the clearest. I was walking through Lochend woods with my dog (6am-ish) and saw, what I initially thought was a dog, appear from the undergrowth at the side of the path ahead (about 30yds) and cross the path to a small stream. I walked further and called my dog so that I could re-attach the lead. At this point the 'dog' turned, and I realised that it was a large cat. It crossed the path again and vanished into undergrowth. I reached the spot of sighting within a few seconds, but the animal had completely disappeared. This makes me think that it was a wild animal and not a domesticated dog.

There was no noise and no interaction. The animal moved quite slowly, and lightly, on its feet." **(Source: *BCIB*).**

Fife
19 reports

11th January: *"I was driving from Dundee to Glenrothes for work on the A92 and passing through a local housing scheme, just past Kilmany looking to my right, there, walking thought a short green grass field, I saw a large black cat, which seemed to be in a hurry. I could not stop the car, as the traffic was very busy in each direction. The cat was longer then any dog and seemed to walk low to the ground. It had a long thin tail, and had short legs."* **(Source: BCIB).**

5th February: The *Fife Courier* reported: *"A CUPAR couple claim they have conclusive proof that an alien big cat is roaming parts of Fife after they were able to catch the animal on camera. It was the second time that British Telecom engineer George Brown and his wife Jill had sighted the animal on the outskirts of the town. Mr Brown (42) and Jill, a hospital worker, along with their three-year-old son, George [...] first spotted the creature in November last year, in a field near Middlefield, on the edges of the town, and near a housing estate. Mr Brown said yesterday, "We were taking our dog for a walk in the same area as before when we saw the animal around 100yds away. It was sitting at first in some straw in the field.*

It got up and started prowling about and it was then that it spotted our family cocker spaniel pet, Charlie. The beast started to run away and Charlie gave chase, but soon gave up after the animal just vanished into some hay bales. However, Jill managed to get a photograph of it and it is the same animal as we saw a couple of months ago. It had a distinctive shorthaired body and was fawn in colour. It was some four ft in length and two ft high with a very dark bushy tail." Unfortunately, the couple were only able to get one photograph as the battery in their digital camera gave up."*

Fife PWLO Mark Maylin emailed BCIB:

"I am a Police Wildlife Crime Officer in Fife and take a keen interest in big cat

sightings in the area. After investigation, I am certain that the animal photographed in the Courier on 8th February 2007was in fact a rag doll cat. This is a large domestic breed from the USA and some are known to reach 4ft long. One had been reported missing by its owners in the Cupar area not long beforehand. Independent sightings in the East Neuk of Fife of a big cat with pointed ears (no mention of tufts), long smooth tail, reddy brown coat and shoulder height similar to a Labrador. It would be easy to dismiss these as a fox but sightings are reported by foresters and gamekeepers who are experienced enough to know what they are talking about." **(Source: *Fife Courier, Mark Maylin & BCIB*).**

2nd March: Sighting at Tentsmuir Forest, Tayport. The cat was black, and its tail was long - almost to the ground and turning up at the end. It was seen for approximately 5 minutes, but the distance is unknown. The witness said: *"The animal was about the width of a road i.e. from my house to my neighbour across the road. I have a Labradoodle and she is the size of a standard poodle – the cat was as big as her, but sleeker and longer.*

My dog was barking and I couldn't understand what was wrong, as she doesn't usually bark as she was doing. I arrived at the clearing where my dog was, and saw this black animal, which - at first glance - I mistook for a large Labrador, seeing only its head to begin with. The cat stood up (my dog was still barking, but wouldn't go too close) looked at me and my dog, then bent and picked something up in its mouth and proceeded to walk away. It had a kill in its mouth, and walked away at an angle to me allowing me to see its tail, and by then to realise that I was looking at a big cat. It stopped and looked at me for a few seconds, then proceeded to saunter on its way to finish eating its kill elsewhere. I reported the sighting to the police and to our local vet. My husband and I went back the next day and found where the cat was lying when my dog disturbed it. There were also several traces of congealed blood from its kill." **(Source: *BCIB*).**

18th March: Sighting by Mr Gary Haggart at 7.20am in Torloisk, Kennoway. (OS map 59 St.Andrews grid ref. 355055). It was whilst he was on a woodland walk on a cold day, with snow on the ground, and starting to snow. Mr. Haggart said: *"The animal was jet-black in appearance, with a long tail, exactly what I expected at maybe around 60cm. I observed it for 30 seconds, from 75-100yds away. Its height was 60cm, and its length, including the tail, was 1.5m.*

I was walking my dog in the snow, early am when I saw something moving up ahead, and it came out of the trees and walked 30yds away from us, and then it turned 90 degrees and crossed track into denser woods. Luckily my dog did not spot it, but I called him back and waited longer then 5 minutes to let it clear.

I was in two minds as to whether to progress or not as I have spent many years in the country. I have stayed in Blair Atholl for 19 years and I know my wildlife, and this was no fox or a deer. Curiosity got the better of me and we set off, got to the scene and the dog started sniffing and laying lots of scent, and then he started to go into the woods, but I called him back. There was some prints, but nothing clear. There was no noise, very little interaction, and it looked like it had a drink out of some laying water." **(Source: *BCIB*).**

25th March: Sighting at Carigluscar Crags Woodland Park. (NT 05316 91105). There was perfect visibility – it was broad daylight – with a light wind. The cat was jet black, with a long, thin tail carried straight out behind as it bounded, the tip flicked upwards. It was seen for about 40 seconds at a distance of 200m. It was the same size as a large dog e.g Great Dane.

The witness said: *"I am a geocacher, we look for little boxes of goodies hidden in the countryside using our GPS units to guide us. I had found one box on top of Craigluscar crags and was heading for another nearby, going cross-country, when I came to the top of a hill at NT 05521 91045. My next target was in a ruined building in wood below the hill so I was taking advantage of the height to see where the paths in the wood went, when my eye caught something run out onto one of the paths.*

It was heading in my direction, and I thought at first it was a large black dog. It was moving fast, and I expected to see an owner emerge from the woods chasing it, but no-one appeared. It reached the corner of the woods and turned through 90 degrees to continue on a path running around the woods. As soon as I saw its profile I realised it was no dog.

It didn't run, it bounded. It didn't tuck its legs in under itself as it ran; its legs were sticking out the back. It carried its tail sticking straight out and the end flicked upwards as it ran. It wasn't chasing anything or running from anything that I could see. It stopped at the corner of the wood, looked around, looked up, and saw me on the skyline. It was a big cat. It bounded off on a path leading diagonally back into the wood. I headed down the hill, but before I could get to the area four people on quad bikes went roaring up the path it had taken.

I took a GPS waypoint of the position where it had stood still, and it was 214m from where I made my observation. I took the path it had arrived on, and I did find a print at NT 05207 91260 (image supplied). I inserted the stylus of my PDA into one of the holes, and whatever it was needs its claws clipped because it was 15-20mm deep. I don't know if the animal I saw actually made this print, but it was just in the right area and looked odd." **(Source: *BCIB*).**

13th July: Sighting by Jan Miller at 20.55hrs on the Standing Stane road at Checkbar Road junction. She said: *"The large cat was drinking rainwater out of a puddle on the left side of the road as I approached in my car. As I got closer, the cat turned and headed across to the right-hand side of the road and took off into the trees. It was black, with no markings that I noticed. The tail was very long; I would say as long as its body, and the cat's height would be approxi-*

mately 2ft to the shoulders, and the body, excluding tail, approximately 3.5 to 4ft. The sighting lasted about a minute and the closest I got was 20ft.

I was driving towards Kirkcaldy along the Standing Stane Rd. It had been raining quite heavily, and the road surface was dark in colour having been soaked repeatedly that day, and there was a lot of surface water at the side of the road. The sky was very cloudy, and as I approached the Checkbar Road, I was suddenly aware of a movement, and the shape of a very large black cat emerged from the road when it spotted my car approaching. The cat had been drinking the rainwater from the puddle on the left-hand side of the road. The cat turned round and ran across the road to the trees on the right-hand side of the road.

I was talking to a neighbour who told me that he plays golf with a man who has seen big cat footprints in the Balgeddie area of Glenrothes, which is about 6 miles as the crow flies from my sighting." **(Source: BCIB).**

16th July: A large black cat spotted near Auchermuchty. **(Source: *Dundee Courier*).**

27th July: Sighting by Chris Daniels at 7.15am in Coaltown of Burnturk, Kettlehill. The cat was black, with a tail as long as the body. Mr. Daniels said: *"I would say about 2 - 3in thick at least, straight, not tapered like a dog's. As the cat jumped across the road in front of me, I would say at least from nose to tip of tail, it was about 8ft long . I am not too sure of the height - about 3ft. The image I have of the cat is like the puma emblem (pumas are not black - puma sports-wear have a lot to answer for - MF)*

I observed the animal for only seconds as it jumped into the road at a distance of about 50yds from me. I was driving my car in Burnturk delivering papers as I do every morning on the same route. I see a lot of different animals, but never anything like this before." **(Source: *BCIB*).**

15th August: A large black cat spotted near Leven.
(Source: *Dundee Courier*).

Mid-August: A large black cat spotted in Glenrothes.
(Source: *Dundee Courier*).

16th August: A large black cat spotted near Leven.
(Source: *Dundee Courier*).

20th August: A large black cat spotted near Pittenweem.
(Source: *Dundee Courier*).

23rd August: A large black cat spotted near Glenrothes.
(Source: *Dundee Courier*).

25th August: A large black cat spotted near Glenrothes.
(Source: *Dundee Courier*).

Late August: A large black cat spotted in Glenrothes.
(Source: *Dundee Courier*).

10th September: A large black cat spotted near Kirkcaldy.
(Source: *Dundee Courier*).

2nd October: Police received a report of a big, black cat at Riverside Park, Glenrothes. **(Source: *Dundee Courier*).**

10th October: The latest sighting in the area was outside the Homebase store in Glenrothes on Wednesday. A retired couple witnessed an animal hunting for food in bushes near the Homecase store in Glenrothes. **(Source: *The Courier*).**

16th November: *"It was around 3 pm in a field at the back of our house in Star of Markinch. It was jet black all over, with small pointed ears and a fairly big head. Its tail was long and black and the animal was the size of a medium to large dog. My husband watched this animal for around half-an-hour, as it was crouching in the grass some 200yds away. He had binoculars, but did not take a photo. It moved off in the general direction of Markinch, and seemed to come from Kennoway direction.*

It came from the field to the left of our property, and stalked about in the field at the back of our house, but mostly crouching in the grass in a couple of different parts of field. Then it got up and walked off across other fields. I thought it was a dog, but it stalked and walked like a cat." **(Source: *BCIB*).**

Invreness-shire
3 reports

2nd February: Highland - first sighting ever recorded on the Isle of Skye / Scotland.
(Source: *BCIB*).

8th July: Heard by two witnesses in the afternoon at Belladrum (OS ref.NH 528 418). The cat was only heard, but not sighted, and paw prints in the vicinity were photographed. They were from a large unusual feral. The sound was heard for approximately 30 seconds from 20ft in thick undergrowth on a sloping hillside.

Mr. Vincent Galusha, one of the witnesses, said: *"I was five or ten steps ahead of my wife as we were walking on a forest trail in sunny, warm weather. I heard a long low growl, and turned around to see a surprised expression on my wife's face. She mouthed the words, 'Did you hear that? What was it?' I mouthed back, shrugging my shoulders, 'I don't know.' We walked slowly and quietly away, picking up our pace as we went. We agreed that it was most likely a big cat. At the top of the hill we spotted a large paw print in the mud and took a photo of it.*

Soon a teenager on a mountain bike came up to us from a transverse path. I pointed out the print to him and said, 'big cat'. He said that could explain why a fox ran out in front of him further down the hill. Continuing on our walk, we spotted several more large prints". **(Source: *BCIB*).**

August: Seen at Lochaber, when a regular visitor to the area was driving along the new stretch of the A830 between Moss of Keppoch and Kinloid.

A 'huge cat the size of a Labrador' ran across the road near the underpass, leapt the fence and ran off towards Craigmore. The witness said: *"It was definitely a cat of some sort, but bigger than any cat I've ever seen, and definitely not a wildcat."* **(Source: *West Word*).**

Lanarkshire
1 report

1st October: Sighting by Ian Hunter at 9.15am. Mr. Hunter said: *"It was on a back road I had never been on before (I was lost, between Bonkle and Shotts Kirk/Salsburgh). The animal was jet-black, and the tail was long and black, like a snake, I suppose, and curved, not quite as long as the body.*

It was the size of a Labrador dog I would say; strong, lithe, and powerful. I observed it for maybe about two or three seconds and it was 20ft away.

I was lost on a road I had never driven before and it was quite bendy, so I was taking it easy. I drove around the corner and something dark seemed to explode off the road and over the fence into the trees and bushes.

I knew it wasn't a fox because of the colour, so I stopped where it had disappeared, and I saw a very large cat, skipping over fallen trees and around bushes before it disappeared into a field. I kept going and stopped further up when I had finally joined a main road, and looked back and noticed how desolate the area was, just miles and miles of moorland." **(Source: *BCIB*).**

Morayshire
2 reports

2nd April: A local woman, who was returning from visiting friends in Dallas, spotted an unusual creature from her car, very close to the road. The woman, who does not want to be named for fear of ridicule, saw the animal near the Dallas turn-off on the outskirts of Rafford. She told her husband, who said that at first he thought she had been seeing things. **(Source: *Forres Gazette*).**

18th July: Sighting at 08.45hrs at Dallas, on the Knockando to Dallas, nearer Dallas just over the Oak Hill. The witness said: *"The cat was shiny black, with no markings that stood out, and it had pointed ears. The tail was long and thin and curled up at the bottom. I*

t was very similar to a cheetah's tail we both thought, and that is the best we can both describe it. The cat would have been the height of a fox and roughly the same length. It was of very athletic build i.e. it seemed to have large hind and front legs with a very small head. It was like a small cheetah but shiny black.

We were in the car driving towards Dallas, when I noticed something cat-like, walking on the road about 100yds in front. When we got about 20yds away it turned and looked our way, and just strolled on to the verge and into the bushes. It didn't seem at all bothered about us in the car; it didn't run or panic, just a took a steady stroll, looked, and then disappeared.

My first thought was a wild cat, but the closer we got I could see it was pure black, rather like a big powerful shorthaired house cat but really slim and athletic. I do believe there are a few of these cats in the wild. But where they are from, I'm not sure.

I know for sure it wasn't panther size so that rules out the big cats for me. I think it's just one of these mysteries you would love to see solved. I wish I had my camera to get a photograph. Would like to see one captured though just to see one close up. But, I do think they are best left alone." **(Source: *BCIB*).**

Nairnshire
2 reports

18th July: *"This is not a sighting. My horse was attacked by something, which I believe could only be a big animal e.g. a cat. The wounds on her body suggest she was jumped on from behind.*

She is a big horse (16.2hh) and has scratches and puncture marks on her back, both her sides and down her two back legs. Her chestnut (woody growth on inside of leg) has been ripped off, and one of her legs is cut fairly badly, and was swollen to twice its size." **(Source: *BCIB*).**

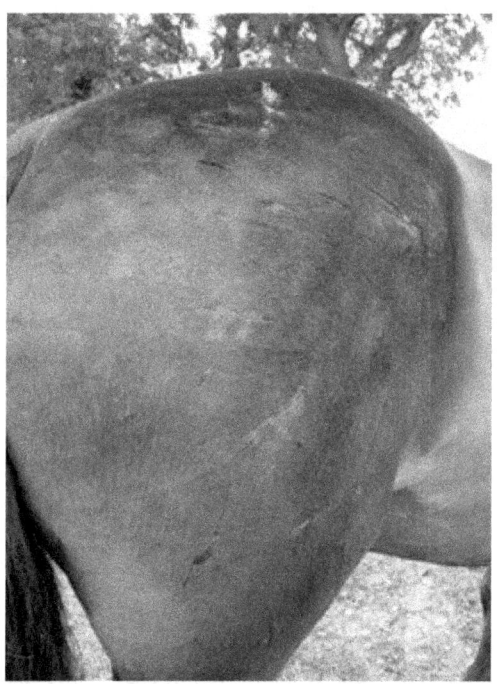

8th August: Sighting of a black, large cat, very sleek and not stocky. Pluscarden. Its tail was long at ⅔ the length of its body, and it was 3ft-ish long by about 1ft tall. The witness said: *"I just came around a corner and the cat darted away from me, and ran up a tree extremely quickly. It was about 100ft from me. I had already sighted this cat about two weeks previously in the same area, but it just ran away at that time."* **(Source: *BCIB*)**.

Perthshire
2 reports

25th May: Sighting by two witnesses at 2245hrs, between Dunkeld and Caputh on the A984. It was black, but they could not see any markings. The tail was long with a u-bend shape. They were not sure about the length of the tail. They observed if for 5 - 10 seconds at a distance of 20m approximately. Its height was about 3ft (size of a small hedgerow) and the length of the cat was approx 7 - 8ft (just short of the width of one side of the road).

Mrs. Lisa Butler, one of the witnesses, said: *"My sister was driving and I was a passenger. We were driving from Dunkeld to Murthly along the A984 a minute or so before the turning for Caputh. The main beam was on as it was 22.45hrs, and very dark apart from the moonlight occasionally breaking through the clouds and the trees. The headlights caught the 'cat' crossing slowly across the other side of the road going into the trees and undergrowth. As the light caught it, it looked at us (almost in defiance!). We were both shocked, but had to turn the main beam off, as there was an oncoming car rounding the bend."* **(Source: BCIB).**

3rd November: Sighting at 2pm of a rusty red coloured animal, like a fox but definitely was not a fox. It was seen in fields beside Glen Lednock Walk near Comrie by two witnesses. One of the witnesses said: *"The beast was at the other side of a large field so it was quite difficult to see its detail. The tail was about a foot long, and the animal was in the stalking position like a cat tracking something; it was about a metre long. It was a large unusual feral, and - as before - we could not ascertain the height, but it was about a metre long.*

We were walking up the track in Glen Lednock and saw a large beast at the other side of a large field. The beast was crouching down as if stalking as a cat does." **(Source: BCIB).**

Renfrewshire
9 reports

8th February: A female witness at Linwood looked out of her back window to watch the deer in the fields. She was amazed to see a large black cat, crouched low stalking the deer. Then it pounced catching a small one by the leg. **(Source: Paisley Gazette).**

March: 12-year-old Daniel Smith and friends spotted what they believe to be a big cat in woods near Houston. He said: *"I saw something big and black walking about 50m away from the footpath and then afterwards, my friends and I saw its paw prints."* **(Source: Paisley Evening Gazette).**

20th March: Port Glasgow window cleaner Gordon McLean was driving home with his workmate when a large black cat ran out in front of them. The drama unfolded when Gordon was out collecting cash from his rounds in Kilmacolm on Tuesday night.
(Greenock Telegraph).

7th April: The witness was driving towards Beith at 23.00hrs. When passing the Howood train station, he spotted crossing the road from left to right a large black cat-like animal. It had emerged from farmland and was only spotted in the headlights for a few moments.

It was definitely cat-like with a long thin tail but much larger than a domestic cat or fox. The witness said that on a night he has seen many animals in his headlights, but nothing like this before. **(Source: BCIB).**

28th April: Sighting at Gourock / Fauld's Park Amazon Factory. The witness was driving to work at 19.00hrs, and as he turned the corner into the car park of the Amazon Plant he saw, *'bounding across the road, a big jet-black panther-like cat.'*

He described it as much bigger than a Labrador. He got a clear view of the rear quarters as it disappeared into bushes. **(Source: *BCIB*).**

17th May: The Greenock Telegraph reported: *"THESE are the sensational images of a big cat claimed to be roaming wild on a Gourock hillside, reports the* Greenock Telegraph.

The Telegraph reported that the images were captured on a mobile phone, but in fact they were first seen on CCTV and the phone footage was taken off that. The cat emerged from bushes and strolled across a car park as Amazon workers turned up at 8.30am.

One witness said: "There were people going in and out of the building, but it was not fazed by anyone."

A spokesman for Amazon UK, where the footage was taken, poured cold water on the cat tale. He said: "There are a couple of cats roaming around, but they're feral cats. Workers at the factory feed them."

(Source: *The Greenock Telegraph*).

25th May: Sighting by Steven Clynes at 12:45pm in Inverclyde, Greenock, Scotland. The exact location was: N 55° 55.570 W 004° 47.648. The cat was brownish red, and was seen for 4 seconds from 500ft.

Mr. Clynes said: *"I was approaching the location of a geocache (which, according to the GPS was 500ft away) and came upon some sheep and lambs who ran at my approach, at the same instant a large cat, which had been lying down near the cache stood up, and bolted in the opposite direction. It was very clear to me that it was a big cat and it appeared to me that it was stalking the sheep.*

This is in the moorland above the Greenock cut near to the path from Overton cottage to cornalees." **(Source: *BCIB*).**

September: Brian Clough of the Paisley Gazette reports: *"A FARMER yesterday claimed a leopard savaged his flock of sheep and killed two of them in a vicious attack.*

Animal carcasses, with their heads ripped off, have been found at the edge of dense woodland; cattle have stampeded, and livestock, which have gone missing, have never been recovered.

Anxious farmer Hugh Caldwell revealed: "I believe a leopard or panther is causing the trouble. And I think there are a few of them around the Paisley area."

He told how three Blackface sheep were attacked during the night and huge teeth holes – spanning nine in – were found on each of their necks. Hugh, who is in his 50s, said: "A dog or fox wouldn't cause that bite span, it's got to be a much larger animal – and a big cat fits the bill."

Hugh added: "There is no doubt in my mind a panther or leopard is on the prowl in the Renfrewshire woodlands. There is possibly a family of them."

He spoke of his fears at the family home in East Mitchelton Farm in Kilbarchan which lies between Howwood and Lochwinnoch. Hugh said three of his sheep were attacked on the same night, adding: "It was the teeth marks on their necks that worried me. The bite marks were from ear to ear.

"All three sheep were alive following the attacks but two of them died after their wounds became infected. The third pulled through and is now back out in the field.

"This is just the latest incident in a spate of them over many months. I have missing animals and I've come across headless carcasses but I can't be sure they are from my flock.

"Dogs and foxes wouldn't eat and strip carcasses of flesh and meat to this extent. But a big cat would do this.

"That's why I'm convinced a leopard or panther is roaming this countryside." Hugh has alerted the police and they are carrying out an investigation.

One officer said: "We are taking this report seriously as we have many others over the years. "If anyone knows anything about a big cat in their area and has photographic proof ring Johnstone Police Office on 01505 404000."

Follow up by Mark Fraser:

"After contacting the farmer, he is more then happy for us to come to the area, set up cameras etc. He is anxious we discover what is killing his sheep. Police told Hugh that they have seen the cat on thermal imaging equipment, and a lynx has been snapped on camera phone. This is ongoing and will be investigated long term by BCIB (Mark Fraser & Brian Murphy). We have the full co-operation from the farmer." **(Source: Brian Clough of the Paisley Gazette and BCIB).**

Early October: A large black panther type cat spotted on East Mitchelton

Farm, Howood by Hugh Caldwell. **(Source: *BCIB*).**

Roxburghshire
1 report

28th December: Sighting at 9:40 pm in Roxburghshire, Gattonside near Melrose of a black cat with the head the size of a human. It had green eyes, and the animal was skulking in roadside bushes as the car approached\passed. The witness said: *"From looking at a few of these sites, I would say it was a black panther - it had a round-ish head as opposed to sleek one."* **(Source: *BCIB*).**

Selkirkshire
1 reports

4th November: Sighting at 11:30 am on Clarilawmoor Farm, Selkirk by two witnesses. The cat was jet black with no visible markings. The tail was long, swooping and around 1ft 6in.

One of the witnesses said: *"I would say the cat was 1.5ft in height and 3ft-3.5ft in length. It was seen from 200 – 150m away for a duration five minutes.*

We were driving along the road, when suddenly the passenger caught the glimpse of the cat on the far away hill. I stopped the car, and the cat continued to walk along the side of some reeds, and when the cat saw me it went into the reeds. I ran half way down the field and made some noise to try and spook the cat out of the reeds. It ran back out of the reeds, along a fence and into the nearby woods. A horse and sheep looked startled by the image of the cat, which never made any noise, or any interactions, with the other animals in the field e. g. horses, sheep." **(Source: *BCIB*).**

Stirlingshire
2 reports

23rd March 2007: *"I thought I would email you a story I was told last week. My mum is a paramedic for the Ambulance Service, and two of her colleagues were out sitting in their car behind the B&Q warehouse in Falkirk on 23rd March 2007. As they were waiting to receive a call out, they saw a huge black panther type cat. My mum's colleagues were a man and woman, and the woman was scared. The man decided to put the car lights on, to get a better look, and the cat ran off into the field nearby.*

They said that he cat was jet black, about 3ft off the ground, and had a long thin tail.

They haven't reported this to any authorities, as they are scared in case they get laughed at. I saw your website, and I thought I would make this comment. I usually go for walks, along that forest area, and I have seen some pretty impressive paw prints recently - I am no expert, they could belong to a dog, or a big house cat, but I have seen some pretty big paw prints there. [...]" **(Source: BCIB).**

19th April: *"I saw a large black big cat in Bannockburn, going down a slope into bushes when walking my dogs. At first I thought it was a dog, but there was no one else around, but dogs don't have ears like that! Then it swept off into denser bushes. Have you heard of anything in the Stirling area?"* **(Source: BCIB).**

Sutherland
1 report

13th June: A Sutherland couple, who claim to have seen a big cat near their Sutherland home, say they have captured the beast on camera. And a copy of a DVD recording of the sighting at Rhiconich, near Kinlochbervie, has been handed over to Northern Constabulary. The witnesses promised to send BCIB a copy of the video tape, but they never did. **(Source: *Press and Journal*).**

West Lothian
3 reports

July 2007: *"Having read an article in West Lothian Courier today regarding the possibility of a big cat being in the West Lothian area, we have a story to tell, but unfortunately not a visual sighting.*

We moved in the Bridge Castle area (between Armadale and Westfield) last July and have two house cats who settled in very quickly. In the past four months we have had three instances where our cats have been terrorised by a cat outside the door to the house.

The first incident was the worst, where our ginger and white male cat had a ferocious fight through a glass pane (cat-flap sized) at the bottom of the door with what we assume to be a black cat, as he then turned on our black and white cat. We think he thought the outside cat had got into the house hence the reason for the attack. This was very traumatising for all of us as they were hissing, spitting, crying at each other for a number of days afterwards, as they did not recognise each other because they had both been sprayed. Eventually they were fine together but with a lot of stress in between.

The second incident happened at the end of June, when it was very apparent that the cat was outside again, as the ginger and white cat was upset, anxious and wary of the room and especially the doorway again. This was not such a violent incident, but it did require the cats to be separated for a couple of days until he had calmed down again.

The third incident happened just a few days ago when we found our male cat hissing and spitting at the door, and there was a very strong smell of cat spray at the door. Again this was a smaller incident, but the cat outside seemed to have sprayed unprovoked as the male cat had been in another room until a few minutes before being found at the door hissing.

All three incidents have happened in the evening - mainly in the dark - and always at the same doorway. We have been spraying cat repellent, but it does not seem to be having the desired effect. We feel that either it is not a local cat or it is a new cat to the neighbourhood as this did not happen until fairly recently and we have lived here since last July.

Unfortunately we have not actually seen the cat concerned, as we would not open the door as the incident is happening for obvious reasons, but it rang bells when we read the article. We thought we would let you know of this and would welcome any advice on deterring the cat in any way." **(Source: BCIB).**

26th July: Sighting at 20:20hrs at Ecclesmachan, near Uphall, opposite Oatridge, College by two witnesses. The animal was black, about 3ft long and moved gracefully like a cat - which first attracted their attention. It was a large unusual feral, seen from 35m away, and watched for 10 seconds tops.

One of the witnesses said: *"We were walking along a path and it crossed in front of us taking no notice of ourselves or my dog! We have no idea what it was, but it was a cat of some sort and definitely not a domestic one as it was far too big!"* **(Source: BCIB).**

27th July: *"I have just finished reading the article in the West Lothian Courier in my lunch break at work and am delighted to see the big cat feature.*

On either the 26th or 27th of July I was taking my children to nursery in Broxburn at around 8.50am. I was travelling along the back road that runs from Pumpherston to Broxburn (as if you were going to the Almondell Country Park) and, as I was approaching the corner where the Tarmac factory is, a huge black cat about the size of a Labrador ran out in front of my car. It was very muscular looking and didn't move like my house cats. It had no markings and was dark in colour with quite smooth looking hair. It had quite a long tail. I am glad I bought the Courier this week, because everyone I have told about this thinks I am mad!" **(Source: BCIB).**

Wigtonshire
1 report

29th September: *"Hello, just to let someone know I had a sighting of a cat, it looked like a panther. We were in the car at the time going between Stranraer and Sandhead on the 29th Sept. It was about 12 in the afternoon and we were driving past a wooded area. My eight-year-old niece was in the back of the car and my partner was driving. My niece and I were looking for deer in all the wooded areas as we were driving past, when we both saw this. It was very large, black and jumped up one of the trees in the woods. I hope this is of some interest to you."* **(Source: BCIB).**

Wales

Anglesey
2 reports

November: (*Two sightings*) *"A friend of mine has sent me a cd he made on big cats in Wales, it was on the radio, but I missed it.*

The area covered is Anglesey, Red Worth Bay, and is the most active part of the island. One lady says she sees a big black cat at least 5 times a year. Another lady describes that in the early hours of the morning she was woken by her stallion kicking the stable door so she went to have a look, and see if he was ok. She looked at him and then walked along the stables and out the door at the end, when 10ft in front of her, was a big black cat. She said it was an absolutely beautiful animal - it showed no aggression to her, but looked nervous. She moved slowly backwards into her tacking room, closed the door, and phoned her house for her dad and sister to come and get her, but by then the cat had gone.

The way she described the cat is the best I have heard, and from such a close distance. She said it was black but had a brown tinge to it and would describe it as being brownie black, sounds like a black leopard.

I have asked him if he can give me their details so I can go and meet them and have a look around." **(Source: Chris Johnston).**

Brecknockshire
1 report

4th February: *"A colleague and I were fell running just to the west of Lord Hereford's Knob/Twmpa, SE of Hay on Wye in the Brecon Beacons National Park. We had cut NE off of the footpath running north along the top of Darren Lwyd, the ridge above and east of Gospel Pass on the Llantony/Hay road. As we approached the col between Lord Hereford's Knob/Twmpa and Rhiw y Fan, we came across a small herd of mountain ponies below the footpath which was now 400m above us (we were on the open fell). About 300yds beyond the ponies was a large reddish brown animal. At first we thought it was a deer (there are no deer in this part of Wales), but when it saw us and started to run away, in profile it was a large cat. It was probably half the size of a small pony, but much lower to the ground with a long tail and small head. As it ran, it stopped to look at us (still about 200-300yds away) a couple of times. Its gait, shape and size meant it was definitely not a fox, deer (or pony!). The ponies (which were not bothered by it - but then they are not bothered by much up there) gave a useful idea of scale and size - the body was probably the size of a large dog, but the legs were shorter, the tail long and head small."* **(Source: *Big Cat Monitors website*).**

Cardiganshire
1 report

27th April: Aberporth. *"We have discovered paw prints in our sand at school and we believe that they are of a big cat and we checked on the internet and they match a picture in the BBC website of the Gloucestershire large cat footprints. My mother has taken pictures on her phone but we cannot download them on to the computer. But we can bluetooth and text them."* **(Source: *BCIB*).**

Carmarthenshire
1 report

Early December: *"I reported it to the police at the time so they would know the date, but I can't remember. 18.00hrs - pale gingery corn colour.*

The cat leapt across the lane in front of the car and was caught in the headlights. It was bigger than a fox and smaller than an Alsatian. I think it was a cougar. Observed for about one minute only 12 ft away. [...] There have been other sightings of it in this area. I think they should be left alone. I think they are living in the old deserted quarry here where there are many caves, and a lake, and plenty of rabbits and pheasant to eat." **(Source: *BCIB*).**

Denbigshire
4 reports

3rd November: Sighting at 15.45 hrs by one person with two dogs in St. Asaph behind HM Stanley Hospital, parallel to the old disused railway track, across the fields from the A55 trunk road of a black cat about the size of a collie or Labrador. It was very sleek, with a long tail, and was seen for about 3 minutes and about 500ft away (maybe less?).

The witness said: *"I came up from the old disused railway track in St. Asaph. It's behind the high school, (Glan Clwyd) and a popular dog-walking track. It runs from Rhyl to Ruthin. I came up through some bushes, and had both my dogs on leads, in case there were any other dogs in the area. We entered the field, which is behind HM Stanley Hospital, and I spotted a black animal over by the fence between two fields, about 500ft away (approx) maybe less? It was black and the size of a collie or Labrador and it walked just like a cat, i.e. from the shoulder. It walked across the field parallel to the fence, then sat and seemed to watch us for about three minutes. I took some pics on my mobile, but they aren't clear unfortunately.*

Although both my very excitable and lively lurchers were looking in that direction, neither barked nor tried to run over, which they would surely have done had it been either another dog or a cat. I would have had trouble holding them back as they are quite big and very strong, and eager to chase cats and play with other dogs. After about three minutes it got up and slunk off towards the lane down which we'd walked.

In a subsequent discussion with my sister, she said about a fortnight before she had been driving down the A55 (just across the fields) and her 13yr old son said he had seen a huge cat in the fields. It was dusk, so about 5.00 pm (these fields are the other side of the lane, I had been walking down. I'm logging this report, which I know cannot be validated, just in case anyone else sees anything and as a warning, I guess.

I notice on the UK big cat website a previous sighting just off the A55 near St. Asaph. I also noticed in the local press (Vale advertiser? Denbighshire Free press) last week, a front-page report on several pets going missing locally.

Since seeing this animal, I have looked at a few websites, and see that N Wales is a bit of a hotspot for sightings. We do have the Welsh Mountain Zoo close by in Colwyn Bay (15 miles away) and maybe something could have escaped and bred?" **(Source: *BCIB*).**

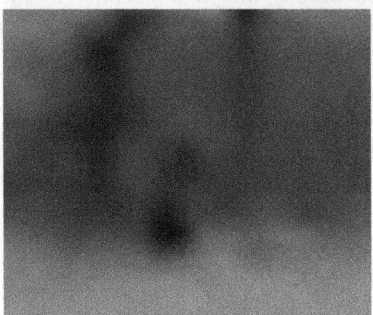

The posts in the photos are 5 ft 4 in high.

April: *"In November last year a gentleman called me from North Wales to say that he had seen a big black cat. I went to meet him, and found a scrape in the wood where the cat was seen. There were droppings underneath the scrape, I think it might have been a female by covering them - all the twigs are piled high.*

In April 07 he called me to say he had seen a big white cat in the same place, while he was on the 'phone to me it was looking at him from the wood. This scrape was different, you can see where all the twigs are pushed to the back and it had urinated in the scrape, the wet area really smelt of big cat. He did get a photo of the white cat, but it was late on in the evening and you cannot make it out. Near Llyn Brenig, just south of Denbigh. There have been sightings in a few areas, Bala is the most active. I think the white cat may have come from there as it has been seen in Bala for the last 4 years.

Around Denbigh seems to be active up to Mold and the Clwydian Range, Blaenau Ffestiniog. A man had two pumas outside his house early one morning, also black cats seen there. Dolgellau, sightings there and Capel Curig and Capel Garmon near Betws - y - Coed but a couple of years ago there. There have been sightings of two black cats walking and rubbing heads, then three months later in the same place a pregnant black cat was seen. Then in the same area three pumas were seen running together.

I know the person who called me tried the Police, the RSPCA, and the Welsh Mountain Zoo, but no one really wanted to know, which is surprising really considering how seriously the Welsh Assembly take this subject. He contacted a friend of mine and he gave him my number as I live near there. These sighting are really very good, the cats where only 30ft from his house. I spent the other weekend driving around all these places and made some really good contacts who are all really keen and said they will let me know of any new sightings and I will let you know if I get any." **(Source: *Christopher Johnston*).**

27th January: *"We were returning to our car, which was parked in the Antelope pub car park, at 20.30hrs when we noticed, in the corner of the car park, a black running creature, which we first assumed must be a fox, but as we were looking and shouting "fox" we both saw the long body and long tail of what we thought was a very large cat. Even in the half-light, it was jet black. Its movement was of a bounding gait rather than a foxtrot.*

As the driver, I was sober and the other witness had just one beer! There was no noise from the creature, but it affected the dog as she was looking and barking in the direction of it walking back to the car up the path. The cat was black all over with a torso of about 1½ ft, and it was the height of a Labrador dog." **(Source: *BCIB*).**

December - ongoing: A series of telephone calls from a gentleman who originally saw and heard a big black cat in the St. Asaph area. He has since been scouring the countryside looking for any signs of big cat activity. Although genuine in his sighting, the gentleman needs advice on what is caused by indigenous or exotic causes. **(Source: *BCIB*).**

Glamorgan
6 reports

6th May: Sighting by three witnesses at 11:50am in West Glamorgan on a mountain above Baglan, Port Talbot, approaching Baglan Reservoir. Paul Sinnott, one of the witnesses, said: *"It had light brown, tan fur, with a very long tail, same colour as above, but with a lighter tip. It was seen for a few seconds as it ran through bushes about 100ft. in front of me. It was at least 3ft high, and at least 5ft long.*

I was walking up to Baglan Reservoir on the mountain above Baglan Estate in Port Talbot with my partner and young son. Just before the sighting a group of people with two large dogs came towards us from the directions of the reservoir then walked past us in the opposite directions. Minutes later we heard a huge screaming noise from the dogs as if they were being attacked (we subsequently saw a man with two Staffordshire bull terriers who I think were fighting with the two dogs who walked past us). It was minutes after this noise that I saw the cat about 100ft or so ahead of me running between a gap in some gorse bushes. I caught a sight of it for a couple of seconds." **(Source: BCIB).**

Early June: An Ystradgynlais man claims to have come face to face with a big cat on the Swansea Valley Mountains. Myrddin Griffiths, of Lluest, says he fled down the side of the Varteg Mountain after encountering the creature. **(Source: South Wales Evening Post).**

3rd August: Witness was driving on the M4 between junctions 45-46 around the Birchgrove area at 17.53hrs. He spotted in a field a large dark coloured cat walking along a wall. It had a very long tail. **(Source: BCIB).**

8th August: Sighting at 11.15 am at Resolven Neath on […] train track by three witnesses. One of the witnesses said: *"It looked like it was a lynx. I say this because I used to work with big cats. It had pointed ears and a long - very long – tail. It was about twice the size of a house cat, and was first seen 150ft away at one point, then I went to about 100ft, and then the cat ran into the hedge. I was walking back from the doctor's and the cat was walking on the train line."* **(Source: BCIB).**

30th August 2007: Sighting today and three months earlier. The witness said: *"It was seen at the bottom of our field next to pine forest. Llanfihangel is the nearest village, but we are between Pont Robert and Llanfihangel. The nearest road is a country road, which I don't think even has a number or classification. The cat was black, with a long tail, and was about 3-4ft - the size of large sheepdog, except the tail was longer than a sheepdog. We have three domestic black cats (one tom is quite big) and when they are in the field they are hard to see. But a cat we have now sighted twice can be easily identified as a cat from a long distance. Peter saw it (three to four months ago) for about 1 minute, and at first thought it was a sheepdog (cat is size of sheepdog) 125 m away.*

Michele saw it today about 250m away for 1-2 minutes in another field next to the forest, but the cat was about 50m further from our house from where the first sighting was. Peter was in the conservatory at 7am having a cup of tea and watching our dogs (3 Jack Russell's) wandering (toileting) towards the pond where the cat was. The conservatory has uninterrupted view of the fields, pond and forest behind. The cat saw the dogs (they did not see cat) and moved slowly away, stopping and looking around to watch the progress of the dogs in a way that showed it was clearly a cat not a sheepdog. We have always had domestic cats at home, and easily recognise their feline movements. As the dogs approached (still very unaware of the cat) it moved further, jumped a fence (as a cat does) stopped, and looked again for the dogs. The dogs came into the pond area so the cat ran off up the field and eventually back into the forest.

Michele was in the field - the dogs wander to 6.30 am - opening up a gate for our horses, when she saw what she thought was a black sheep dog herding/ hunting sheep. As she watched it, it was clear that it was not a dog, but a very large cat. She saw it from 220-250m, and yelled, and screamed at it. It ran into the forest next to our farm. Our horses (six) have been acting strange lately. A number of sheep have been killed in the last weeks; some from fly infestation, but such infestation could be also due to stress to sheep from a large cat. Many of the dead sheep have been half eaten. Local farmer (who owns the sheep on our land) only last weekend took them for back inspection and one had claw or teeth marks on its neck. Peter's sighting was thought to have been so strange it was dismissed (he wears glasses and was told to have his eyes checked).

Michele has now also seen the cat in virtually the same place. We have enquired with neighbours, and our local policeman, and there have been a number of single and isolated sightings by individuals in the area, but none have been accepted or believed. We understand a sighting was made of a large black cat at Dolanog, which is only a mile from us. The area next to the field, where the sightings were made, is forestry and a large cat could easily shelter there without being disturbed. The forest connects with other wooded land, and open farm fields, making a possible corridor to Lake Vyrnwy and its huge wooded catchment area. We know of fallow dear escaping from Powys Castle and wonder if a large cat may have escaped from captivity as well." **(Source: *BCIB*)**.

8th September: *"I had a message left on my answering machine - someone has reported seeing a big cat in Wales. They where driving along the road, and the cat jumped right over the top of the car. His voice was not clear on the machine, and I could not make out all what he said. I have called back, but got no answer, and will call later on and let you know. The area where the cat jumped over the car is called Pont Robert. It is a mile from the recent sighting at Llanfyllin. I am still waiting to hear from the radio station to see if I can get the details of the person who this happened to."* **(Source: *Christopher Johnston*).**

Monmouthshire
3 reports

14th November approx: A large black "mountain cat" has been spotted prowling the hills overlooking Waunlwyd. Father-and-son Peter and Gareth Whittle caught sight of the unusual beast at 11.40am. As they crossed the cattle grid outside Cwm cemetery the big cat walked out in front of them. Gareth said: *"There was no mistaking it – it was about 4ft in length, black and had a big tail. It was* very, very agile, but the head was quite small so I think it would have been a female. It was big, too, not as big as a lion or panther, but it was still very big. It was one big lump of a feline." **(Source:** *http://icwales.icnetwork.co.uk*)**.**

21st November approx: Denise Selway spotted a large black cat while walking her dog George not far from her home in Hillside Terrace, Waunlwyd. **(Source:** *http://icwales.icnetwork.co.uk*)**.**

Month Unknown: Sighting of a black, smooth coat. The witness said: *"We could not see the tail as it was sitting and had its back to us. A friend and I were walking over Bangor Mountain as it was a nice clear day. We were walking down past a farmhouse towards a river, which flowed over the road. We were talking when we both stopped and turned our heads (looking to the left). In the field was a large black shape; we were confused at first and thought it was a tree trunk. We laughed, but then it turned its head to the side and we saw the shape of its head; it was definitely some type of large cat. We stared at it, frozen for at least one minute, then we ran. My friend said that it had chased after us, but then disappeared. We were so paranoid after the sighting that we were stuck in the middle of nowhere, and thought we saw something everywhere we looked. Eventually we got back to civilisation; we didn't report it as we thought we would just be laughed at, but my friend and I did submit our experience on the BBC website."* **(Source:** *BCIB*)**.**

Montgomeryshire
1 report

25th May: Sighting took place near Minervan. Witness spotted a large black cat the size of a "large sheep dog." It was daylight, the animal was on a grassy hill 200 - 300yds away from the witness. The cat noticed him, turned to look for a few moments, before turning back and running away towards farmland. Described as having "quite a thick tail." **(Source:** *BCIB*)**.**

N Ireland

CO Antrim
2 reports

11th May: Stephen Philpott was informed, by the police, of an attack on a dog by a puma. The dog chased the puma into the bushes, the witness heard a bit of a "melee", then the dog ran out again with the puma following. As soon as the puma saw the owner it stopped, turned and ran back into the bush. The dog suffered severe facial injuries.

Exact location not known (may not be Antrim) **(Source: *Charlie McGuinness BCIB*).**

December: Police have ruled out a 'big cat' as being responsible for attacks on sheep in the Ballymoney area. Recently the *Times* reported on growing speculation in the district that the so-called 'Beast of Ballybogey', which created a media flurry several years ago, may have been back on the prowl after several sheep were killed on land owned by Ballymoney DUP councillor Cecil Cousley. This week, however, police moved to rule out the big cat theory after they worked closely with Ballymoney Council official Karen Mitchell. Sergeant Fiona Stirling of Ballymoney PSNI said tests were actually carried out on two dogs, which proved they were responsible for the sheep attacks. **(Source: *Ballymoney Times*).**

CO Armagh
1 report

25th January: A panther is believed to be on the loose in south Armagh. *The Sunday Times* reported at the weekend how a large cat has been sighted by scores of witnesses including a senior member of the PSNI. It was also reported that the panther had caused damage to cars and properties. It was suggested that the panther has set up home in a woodland area between Markethill and Poyntzpass. **(Source: *Newry Democrat*).**

CO Derry
1 report

May: A family says they are living in fear because of attacks on them and their pets by a wild cat. David Smyth trapped the cat outside his home in Burntollet, Co Derry. He later released it in countryside 40 miles away. But about 60 hours later, he found it sitting outside his back door.

Mr. Smyth said: *"We moved in here last March and since then the large orange and white coloured wild cat has attacked my sister Patricia and it has badly injured our pet cats Mister and Missy.*

The neighbours say this wild cat has been here for years. Recently I had to leave the house door open because we were expecting a plumber to call. The wild cat got into the house and attacked Mister and Missy. There was blood everywhere, but thankfully they survived. However, they're afraid to go outside now.

We can't leave any doors or windows open. I contacted an animal shelter and they loaned me a trap, which I used to trap the wild cat. I then brought it to a vet to have it neutered.

I then drove 40 miles away to countryside between Coleraine and Kilrea and released it thinking it would stay there, but I got a shock at the weekend when I opened the back door and there it was again. People have told me to have it shot or poisoned, but I don't want to do that, I don't want to be inhumane. I'll try to trap it again and next time I'll take it to Scotland and release it there, keeping a body of water between it and us." **(Source: *The Independent*).**

CO Tyrone
1 report

27th March: Featured in the Tyrone Herald on Monday past, an amateur photographer and his brother - Ryan and Aiden Bradley - recently happened upon a puma-like monster cat in a field in an area commonly known as the 'Fort' at Ballygowans, close to Omagh town. The duo, which was testing out a new camera at the time, managed to take a photo of the beast before it made off over a nearby hill. The local men say they are in no doubt about what it was they saw and yet despite this and notwithstanding the photographic evidence, nothing is to be done by the authorities. **(Source: *Tyrone Herald*).**

Sitting along the tree-line. This photo was taken by local brothers.

1st February: Residents of the Clogher Valley area have been warned by police to be on the lookout for a large cat-like animal, which was spotted by an Aughnacloy man on Sunday evening. The unidentified creature was seen on the Favour Royal Road at about 10pm. Members of the public are asked to be vigilant and any sightings of the animal should be reported to the police. It is believed to be the first appearance of the so-called 'border beast' for a number of years. **(Source: *Tyrone Times*).**

Ireland

CO Donegal
2 reports

10th June: Gardai have mounted a surveillance operation after reported sightings of a puma on the outskirts of a Donegal village.

A local man in Newtowncunningham, a village close to the Derry border, reported two separate sightings of a suspected large wild cat over the weekend.

Paw prints found at the scene close to the N13 have been identified as belonging to a dog, but gardai say they are taking the reported sighting seriously. Gardai were first alerted to the possibility of a wild cat roaming in the area on Sunday morning but their surveillance operation has so far yielded nothing.

A senior garda officer confirmed that images of paw prints found at the scene had been emailed to Dublin Zoo, but experts there confirmed they were canine. He said: *"We were alerted by a local man, who knows quite a bit about animals, about a brown animal with a bushy tail bigger than a fox."* (**Source:** *Anita Guidera*).

3rd June: A witness reported seeing "a cat-like animal, the size of a labrador", near the village of Burt about a week ago. He said that it was "the rare way it took off", that attracted his attention. (**Source:** *Anita Guidera*).

CO Monaghan
6 reports

Approx 4th June: A large black cat sighted around the Emmyvale area. (**Source:** *Charlie McGuinness*).

Approx 28th May: A large black cat sighted around the Emmyvale area. (**Source:** *Charlie McGuinness*)

27th May: A wedding party saw a large black cat and apparently many of them snapped away and caught the animal on film. (**Source:** *Charlie McGuinness*).

10th May: A sighting of a large black cat by Northern Sound Radio presenter Joe Finnigan in Valley Bay, South Monaghan.

He is not quite sure of what they saw. It had a long tail, but seemed to be definitely feline. Chris Moiser did a ten minute radio interview on Northern Sound and found the witness to be genuine; also several members of the republic rang in to report their sightings, although they wished to remain anonymous.

Charlie McGuinness visited the area the next morning and found the cattle in the field to be acting in an agitated manner, all huddled in the corner and very skittish. In fact he later discovered that they had broken away from their original field, over a 2ft high wall, barbed wire, smashing a post as they did so, across a crossroads and through another field - a distance of half-a-mile. There was also an area of flattened grass which looked as though an animal had been laid watching the cattle. (**Source:** *Charlie McGuinness BCIB*).

5th May : Sighting at 19.50hrs in Tully at the rear of Tully House of a large, jet-black, 3ft long cat, with a 3ft tail. It was seen by the sister of Charlie McGuinness. Two other people were present, but only one other saw the animal. The cat used the exact same route as the animal videoed by Charlie in June 2004 - Charlie had been stood looking at the area only a few minutes earlier, but must have come inside just before the cat came out. **(Source:** *Charlie McGuinness*).

Charlie McGuinness has informed BCIB of another sighting in Monaghan from January - but sticking to time honoured tradition Mark Fraser lost the paper he logged the details on - in his defence he was at work at the time. We shall publish the sighting when we get the details again

CO Sligo
1 report

January: Sighting by Sandy Perceval at Derreen, Ballinacarrow in Ballymote. He has seen many over 40 years on the Temple House Estate, Ballymote, Co. Sligo (see woods round Temple House Lake on map). It was a large unusual feral. They are usually black but […] tabby and one smoke-grey 20 years ago. The last sighting was in January 07 of a dark brown animal with tufts, which are unusual. It had pointed ears. Some cats are bushier than others are, so are their tails. The tail is roughly half the body length. The last cat was seen during a woodcock shoot by several people. He said: *"I've seen them many times, but never know when they may appear."* **(Source:** *BCIB*).

USA Round-up

(A few snippets - for the full round up visit www.bigcatsinbritain.org)

Alex Mistretta, from Chicago, is BCIB representative in the USA. He has been fascinated by cryptozoology for as long as he can remember.

He graduated from the University of Illinois at Chicago, with a BS in Psychology and a BS in Anthropology. His choice of Anthropology was motivated by the interest in Cryptozoology.

Mystery Black Felid Photo: Identified - Florida: The picture was taken in December 2005 by a retired biology professor from a Georgia university, Dr. Edward Yeargers.

The professor wrote in a letter to Loren Coleman *"I was a biology professor at Georgia Tech for 30 years - I'm a close observer. This one was about 18. tall when it sat on its haunches (much taller than a domestic cat), had pointed ears and a short tail. When it walked, its hindquarters were higher than its front. I have also seen bobcats with conventional colouring in my yard, so I know the habitat supports them. Unfortunately, the whole area is about to be cleared and developed."* **(Source: *Loren Coleman - Cryptomundo*).**

January: A forester was chased into the Chattooga River by a 7-foot-long panther with "jet black" fur. Terrance Fletcher, a technician with the U.S. Forest Service, dove into the frigid water and crawled up the bank in South Carolina to escape.

He said: *"The animal started running, so I decided to run and get away and jump in the river to get across to the other side. It was a life-changing event for me."* **(Source: *The State, South Carolina*).**

February: A large black cat seen ripping guttering off a house in Port Charlotte, Kingsgate. (Source: *Charlotte Sun Herald*).

March: *"Various livestock have been found mutilated in the last few months in Raintree Lakes, Upshur County, Eastern Texas. One farmer found one of his calves with both stomach and throat ripped open. Mithchell Bransford had a close encounter with a black panther in April, which some residents feel has been responsible for the mutilations.*

He was unable to identify the species, but he told wildlife experts that it was ei-

ther a black panther or black jaguar. I am not sure if by panther he meant leopard or mountain lion. Local authorities stated that people were seeing either black otters or black hogs and misidentifying them as large cats.

Nearby Cherokee County also had its share of sightings in March and April. Eyewitnesses in Dialville claim to have seen a large Black Panther with tree cubs on multiple occasions." **(Source: *Alex Mistretta*).**

April & May: Over 30 residents called the local police station in Vineland, New Jersey, to report sightings of a black panther. This was preceded by a sighting at about 150yds from Zoe Paraskevas on Saturday April 28th. Zoe managed to photograph the animal at that distance. He estimated the animal to be about 80 pounds and the size of a German shepherd. Some of his neighbours have reported hearing growls near their houses. Animal control officers claim that an overweight house cat was responsible. **(Source: *Alex Mistretta*).**

June: On June 3rd in Wiscasset Maine, farmer Lee Straw woke up to find 15 of his sheep slaughtered. On June 4th, Lee Straw woke up to find 14 more sheep dead in the same manner. Out of the 29 dead animals, only two had signs of having being eaten. Coyotes were ruled out because they kill for food. Wiscasset is located on the Midcoast of Maine, and more precisely in Lincoln County. Lincoln County, along with nearby Waldo County are known for their black panthers sightings, leading some to suggest that black panthers may have been responsible.

June: On the 29th of June the *Lewiston Sun Journal* posted a picture of a large cat taken in Sidney, central Maine. The newspaper speculated that the cat might be a mountain lion. The author of the photograph wishes to remain anonymous.

Officially, Maine hasn't had a mountain lion population since the late 1800s; though a mountain lion was killed in the state in 1938. And since then, aside from eyewitness sightings, definite proof of a mountain lion population has remained elusive. Fur was recovered and identified as belonging to a mountain lion in the mid 1990s, and in 2001 some tracks were identified as also belonging to a mountain lion. This of course doesn't prove that a population exist, as these animals could have been released into the wild by certain individuals. Unfortunately, the tail is hidden by a tree in the current photograph. Tail length and thickness often help in specie identification. Mark Latti a spokesman for the Department of Inland Fisheries & Wildlife said that a biologist was dispatched to look for tracks, and measure the rock upon witch the animal was standing in order to obtain a better sense of scale. He also stated that it is very unlikely that a population of mountain lion exists in the state.

Five days later, on the 4th of July, the *Blethen Maine News Service* ran a story of another sighting, this time in Oakland Maine. Oakland is less than 15 miles from Sidney. The eyewitness, a Kelvin Higgins, decided to tell his story after he read about the Sidney sighting. The Higgins sighting took place in April of this year.

Higgins saw the animal from his porch. The cat, about 30yds away, was grooming itself on a rock. The rock was where Higgins found a little bit of fur and skin. According to Higgins, when the cat saw him, he casually stood up, stretched and bounded off. Higgins cannot say for sure if it was a mountain lion, but he described the cat as 4ft long, with an equal length tail and an estimated weight of 100lbs. The fur sample was sent to Southern Illinois University for analysis.

UPDATE:

The fur sample collected by Kelvin Higgins in Oakland, Maine from his April sighting turned out to be from a red fox, according to DNA analysis. Higgins does not dispute the result, but is still adamant that what he saw was a large cat. **(Source:** *Alex Mistretta***).**

October: *"I saw an enormous cat in a wooded area in Kansas City, KS on Sunday, 6th October 2007. The cat passed behind a home then disappeared into the woods.*

I stopped at the home yesterday (6th December 2007) and I told the resident that I had seen an enormous cat bigger than my 85lb. German shephard dog, and I told him it passed right behind his house and went into the woods. To my amazement, the resident told me he sees the cat periodically in the same place. He said he thought it was probably a cougar. However, that cat is definitely NOT a cougar. It looked more like a large, shorthaired house cat. The cat is very yellow (no spots). The tail appeared to be short, like a bobcat. However, it had a very wide, muscled body, making me think of a lion. The fur was short and

slightly "fuzzy," like a house cat. I am extremely intrigued by that cat, and I think it bears investigaing."
(Source: *BCIB*).

Melanistic Bobcat Caught in Florida (August 2007) *I received this report by email about a captured melanistic bobcat from nuisance wildlife trapper H. who has graciously allowed me to post his report as it is.* "I am a state licensed nuisance wildlife trapper located in southern Florida. I thought you might be interested in this. On Wednesday night, I was contacted by a very excited/scared landowner in Martin County, Florida. The day before something killed her 30lb male turkey. She arrived home Wednesday evening to begin searching for the carcass, which she couldn't find on Tuesday. As she came out of the cane grass she saw what she described to me as a large black cat. Larger than a bobcat, more like panther size. She was convinced that what she saw was someone's escaped black jaguar or black leopard. I headed up the next morning to search for sign of a large cat but found only bobcat size prints along the bank of the Okeechobee Waterway (St. Lucie Canal) on her property known as "The Last Stand". After doing some research, I believed that what I was searching for was a melanistic bobcat.

I set up the live-catch trap baited with a fresh killed rooster. The next morning I got the cat. A 16lb 2-year-old, healthy male melanistic bobcat. The cat is alive and well, and currently being kept at a local wildlife center until a decision is made as to where he will end up. I have been able to track down 13 previous confirmed (either captured cats or photo evidence) records of melanistic bobcats in south and south central Florida since the first one in 1939. I believe this male makes 14 and currently the only living one in captivity. I have been searching the internet trying to find people to e-mail who might be interested in hearing of this find."

September: Workers at the MHF Packing Solutions in the Sweetwater Industrial Park, Tennessee, saw some type of big black cat on a hill behind the plant around 8:30.

"We see deer out here all the time," said Charlotte Shell. But this was different, workers said. Shell said some of her co-workers first saw the big cat as they walked past an open warehouse door. Shell took pictures of the cat that perched by some brush on the hilltop and was still there meandering in and out of the brush some two hours later. **(Source:** *The Advocate & Democrat*)**.**

October: While not as numerous or frequent as sightings of black panthers, reports of large maned mystery cats occasionally surface from various areas of the United States and Canada. The latest is from West Virginia where Jim Shortridge, a bow hunter, had a sighting of an animal he described as an African lion.

Jim Shortridge owns some 40 acres of land near the Cold Knob Mountains in Greenbrier County. He was at his hunting shack, on October 17th, when he heard a growl. He couldn't make out what kind of animal was growling, but - a little unnerved - he yelled at the animal which initially ran away. Thinking the animal gone, Shortridge walked over to his truck to retrieve his bow.

He then returned to his hunting shack, and shortly thereafter the animal returned as well. It was still dark and only 5:30 am, so Shortridge grabbed a flashlight to get a better look at the animal that was now only about 10 to 15yds away. He described it as looking like an African lion with a mane, and he estimated the weight at between 250 and 350lbs. The cat paced back and forth, growling more aggressively, as the light from the flashlight seemed to irritate the animal. I went through several sources for this story and none mentioned how the animal departed, but a couple articles stated that the sighting lasted for about 40 minutes. In any case, even in the dark, Shortridge had ample time to obtain a clear visual of the animal.

Curtis Taylor, chief of the DNR's Wildlife Resources (Division of Natural Resources) stated that Shortridge's report was the second that they has received recently. No detail was given on the other report. Robert Mclung, an animal control officer for Greenbrier County, is also taking the report seriously. McLung, along with Shortridge, are leaving chicken meat in the area alongside motion-triggered cameras hoping to photograph the animal.

Taylor and Mclung both feel that the animal was most likely abandoned by its owner for whatever reason. I have to agree that for now this is the most likely scenario. Perhaps the animal was kept as a pet and released when he became too large and feeding became inconvenient. If Shortridge's weight estimate is accurate, between 250 and 350lbs, the animal could be a fairly young lion. (*Alex Mistretta*).

"Some interesting comments were made recently by Jim Boggs, an assistant field supervisor with the U.S Fish and Wildlife Services in Lafayette Louisiana regarding black panther sightings in Morehouse Parish, Louisiana. What I felt was of interest was his comments on specie identification for these sightings, which exemplifies the difficulty in dealing with specie identification for black panther sightings in the United States in general.

Officially Morehouse Parish, Louisiana is not home to a native large cat. Some have suggested the Florida panther as a possibility, but Boggs feels that if these animals exist, Texas cougars are a more likely possibility. Florida panthers are endangered and are not known to breed outside of Florida today. Their historic range however did include Louisiana as well as Texas, Arkansas, Georgia, Mississippi, Alabama, Tennessee and South Carolina. While Boggs does point out correctly the difficulty in a Florida specie making the trek today across the Mississippi, he doesn't address the possibility that perhaps the Florida panther has never left Louisiana, or/and some of the neighboring states. The Texas cougar is not endangered, and Boggs theorizes that a few animals from southeast Texas could have moved into Louisiana in order to escape growing human encroachment.

The Florida versus Texas subspecies debate may be a moot point however, since many of the sightings involve a black animal, and neither of these is known to exhibit melanism The only theories Boggs offers here is that perhaps these are reports of jaguars which are known to exhibit melanism, or perhaps a jaguarundi. Both of these live in Central America today and the jaguarundi has been reported in Texas on occasion. The jaguar hypothesis is intriguing and nothing new, since it once thrived in parts of the southern United States.

One sighting by resident Tammi Gardner includes one larger cat with cubs, perhaps indicating that these animals are breeding in the area. Boggs appears somewhat sceptical that a breeding population exists and mentions that the animal seen by Gardner may have escaped captivity and bred with a male bobcat. Black bobcats do occur as demonstrated by the picture taken in Palm City, Florida that can be seen on the BCIB website." (***Alex Mistretta***).

Picture courtesy of John Finch

Have you seen a Big Cat: Please report to us at BCIB. We will respect any confidentiality and will not even make your report public if that is your wish. We are, primarily, researchers.

Date: Location: description of the cat: colouring, length of tail, shape of ears etc. Try and include as many details as you can remember. Was there any physical evidence left behind. Photographs.

www.bigcatsinbritain.org
Mark Fraser 35 South Dean Road, Kilmarnock, KA3 7RD, Ayrshire, Scotland - bigcatsinbritain@btinternet.com

----

Big Cats in Britain
Investigation throughout the nation........................
by Mark Fraser

A Cat is a Cat, is a Cat!

The Big Cats in Britain group (BCIB) came about by accident, it certainly was not my intention to create such a group, far from it. Now it has grown into a nationwide, and yes, even worldwide, big cat research centre. Di Francis can claim the credit for that whether she likes it or not. It was her books in the early 80s that inspired me to take up the quest, which I must admit, has become an obsession.

It was in the early 1980s when I began to search for evidence around my home area of East Yorkshire. I met many witnesses and made many mistakes, more importantly I made many new friends, and learned a lot.

In 1999, I decided to try and bring researchers together to share information and help build a bigger a picture of what might be occurring. I was not the first to try this, but was the first to succeed, in a fashion! This took the form of an internet mailing group, and it grew from there. Gone are the days of letter writing and typewriters! However, for various reasons it is not possible to bring everyone together. The BCIB prides itself on teamwork, facts, honesty and friendship. We are also very aware that nothing we do now is new; it has all been done before.

BCIB consists of people with one common aim, which is to discover exactly what the species of cats roaming the British Isles are. There is no doubt that they are here, and have been for a long time. The questions are: What are they? How long have they been here? How many are here? And why do the facts, when looked at, indicate an unidentified species? Although some group members may disagree with me on this point.

The investigation of big cats is our primary concern. Also the confidentiality of our witnesses' personal details, and location of sighting, if that is their wish.

Members of the BCIB each have their own opinions and ways of working, and we may not always agree with each other on everything. That we can disagree, and work towards a common goal, is one of the strengths of this group. We are based in several parts of the country, and have active contacts the length and breadth of the British Isles.

BCIB is open to membership, and encourages it. At the time of writing we have members from all over the British Isles, the Irish Republic, Australia, USA, France, with new people joining all the time. The group welcomes members who have differing interests such as, remaining passive and just browsing our website, or news gathering, attending and arranging vigils. Each has an important role to play. And we keep each other focused when we might be tempted to go off track! Members of BCIB include researchers, authors, police officers, scientists, zoologists, environmentalists, soldiers, housewives, truck drivers and others.

We are also associated with other research groups, most notably the Rutland & Leicestershire Pantherwatch Group run by father and son team, David and Nigel Spencer. They have provided extensive information on large cats in and around the Leicestershire area for the last ten years, have made many positive contacts with PWLOs, zoos and land workers.

Our aims and objectives are:
To gather as much evidence and information on the British big cat as possible.
To compile a county-by-county sightings register
To investigate on-site whenever possible.
To discover exactly what species of cat is roaming the British countryside.
To discover how they came to be present or, if indeed, they are an indigenous species
To respect witnesses' confidentiality at all times
To collate the biggest online archive on British big cats
To take conclusive video or photographic evidence that cannot be disputed.
To open a small 'museum' displaying the information that the group has.

Our field research concentrates on attempting to obtain conclusive video or photographic evidence that cannot be disputed. Photographs of livestock kills, prints, as well as big cats themselves - all are most welcome, as is video footage. All evidence submitted will be returned.

At present, my main area of activity is with vigils - maintaining a watch with a hide, with photographic and video equipment - although I still enjoy meeting the witnesses. Some people think vigils are a waste of time, and say "*I cannot see the point of it*" or "*you'll never see one.*" In fact, I have had three positive, clear sightings while doing such vigils which rather knocks those comments off the shelf somewhat. I would not have had these sightings had I not been out there, it's as simple as that. These are real flesh and blood creatures that need three things: food, water and a mate. Ideally we need a permanent hide for at least three months if we can realistically hope to achieve positive results. But with modern day living, work and family life this proves to be impossible to achieve. The only way we can do this is by a rota system, and a whole load of people ready to brave the elements. Are you one of those people?

We have talked to very many of witnesses all over the country. Respect is always given to a contact, and we always follow any instructions given by them such as confidentiality, and keeping the sighting a secret. We have been called in by farmers, factory sites, the police, and the MOD, to name just a few, in a bid to discover what animal is roaming their particular stretch of land. Our core of experts in different fields are all on hand to offer advice.

BCIB have their own website at *www.bigcatsinbritain.org* or *www.scottishbigcats.org* with a member's area which displays all the latest sightings. We have a vast archive of news clippings, which are available to all to browse at their leisure. There are sections on many aspects of the subject for you to peruse and hopefully add to. And you are welcome to add any comments you wish to make. At present there are over three thousand pages on the website and it continues to grow.

BCIB have taken part in several TV documentaries and shows. There are several more in the pipeline to watch out for. We are more than happy to assist any radio, television company or journalist with their research into this subject. What we will not do is make up reports, or fake evidence, for the sake of a story. We do not make any outrageous claims, or falsify reports in anyway, nor will we pass on witnesses' details without their permission. We are searching for the truth, however dull, boring or exciting that may be.

BCIB hold social events each year with the main item being the conference. We always have a great line up of speakers for your enjoyment, and we give you the chance to ask them any questions you wish. The event is open to non-members too. It is held in different parts of the country each year. We discuss and debate in a friendly and open manner to help further our knowledge and understanding on a subject that, for some of us, is a way of life.

Our yearbook is very successful, and it contains a round up of sightings, articles and major events that have taken place.

Members share their sightings, and can also access the national database. They follow up sightings, conduct field investigations and night vigils.

Membership is £10 a year (at the time of writing) with special membership packs becoming available in the near future, benefits include:

- Access to the members area.
- A personal webpage!!
- Six newsletters a year keeping you up to date on what the group has been doing
- Access to the members only mailing list where you will be constantly updated on what is happening when and where.
- Updated information of sightings in your area, along with the chance of investigating locally on our behalf, and more importantly with our back-up
- Listed on the website as an investigator for your area
- A chance to attend the vigils we hold every year in the hope of finding evidence.
- Attend social events
- Become a member of the biggest and most active big cat research group in the UK.

The main artery of the group is the witnesses - without you we simply would not exist. I commend the people for taking the time and trouble to tell us of their encounter, an encounter that many people did not ask for, or particularly want. Many former sceptics have been converted by their chance meeting and become active members in the field of British Big Cat research. We are appealing to you to report any sightings that you may have had, no matter how trivial you may think them. There is a form and an address included at the end of this chapter, which you use for reference.

Photographic evidence and video footage are very important, and we ask you to report any such material to the BCIB, no matter how long ago it was taken. We can copy your footage on to DVD or video tape for you and even make extra copies.

The BCIB receives reports of shootings by farmers and many road kills. However, what is essential, is *retrieving* that body. BCIB is not a money-making organisation, but good photographs, video footage, and especially a body, can make the photographer or finder some remuneration. All we ask is that you report any such occurrences to us in the first instance. We want nothing other then to gain this evidence, and with our extensive contacts BCIB will make sure that any money earned will go straight to you. But we need that evidence!

One word of caution that we will offer to the witnesses concerns the finding of prints: Many people have a genuine sighting but in their enthusiasm (and understandably so) they look for evidence and may find a print - usually its a dog print no matter how big it is; this will detract from the sighting and make others want to dismiss it as a dog. Be very careful with prints before you approach the press. We will be happy to tell you what animal made the print first, to save frustration in the future. But we do ask that you will not be disappointed if the print is identified as that of a dog. Do note if the print shows claws then it is most probably not a cat; prints look bigger in mud and tend to display and in this case, size isn't everything!

Report A Sighting

Your report is much valued, even if you have reported it to others in the past and no matter how trivial you think it to be. Your experience will help us estimate the numbers of cats, their territories and habits etc. All data except your personal details will be added to a database. Sightings today are a common occurrence you are not alone. A report of a big cat in the British Isles and Ireland average three a day. We are always interested to hear from you.

Witness Comments

Thank you, I have been very impressed by the courtesy shown by your members and I will be vigilant when I am on my travels as I feel it is only a matter of time before there is some sort of incident.

It sounds like a well thought out and professional organisation. I'm happy to share more but sadly no photographs yet. I caught a possible sighting again this evening of a dark catlike movement but at night time with night vision equipment. Is there any funding yet into research into big cats in Scotland? Definitely interested in your conference and I would also be keen to see any field reports from the Scottish Borders etc.

It is over 40 years since the first public spate of sightings hit the headlines with the Surrey Puma, yet we are still no closer to solving the mystery. Hybrids? or a relic, indigenous species that we never knew existed alongside us, ever since the Ice Age?

BCIB average three sightings daily from all over the countryside including Ireland, in fact people these days are more likely to see a big cat rather than a pig. Large unidentified felines are here in the British countryside, that much is blatantly obvious to all but the most blinkered.

In the past there have been pumas and lynx caught and shot and instances of leopard and other jungle cats run over. However the true British mystery big black cat, has long eluded capture. It is now time to progress with this mystery and find out conclusively what these cats are. We cannot do it alone, report your sighting, your news, your video footage and pictures along with any other hard evidence that you may come across is highly valuable.

The BCIB are planning to open a museum, *(an idea first put forward by Di Francis in the mid 90s)*. This is in order to display any artefacts which we have and may have in the future. Sightings and news reports will also be available on microfilm.

Big Cats in Britain - The Magazine

March will see the launch of the first issue of the groups new quarterly magazine simply titled '*Big Cats in Britain*'.

It has an editorial team of seven members with advisors such as Di Francis we can call upon. 50 pages A5 in size.

It will concentrate solely on the Mystery Big Cats of Britain and will also include sections covering various mystery cats around the world.

BIG CAT YEAR BOOK 2008

Witness Credits:

Quincy00, Cactus Flapjack, Chris Kerfoot, David Gater, Ian Duffin, Patricia Lewis, Geoff and Sylvia Killgallon, David Zaborowski, Joy Knowles, Trevor Smith, Suzanne Young ,Matthew Tucker, John Almond, Raymond Franklin, John Page ,Anne Raiment ,Terry Brown,

Terry Foster, Phil Hodgson, Paul Twigg, Gemma Prest, Brian Wardle, Robert Chapplow, Paul Ryder, Eddie Davies, Elizabeth Mitchell, Gill Bunker, Kevin Wright ,Peter Roberts, Paul Caffrey,

Andy & Sarah (Cornwall), Stuart Buxton, Lalla Blackden, Kim Redgrave, Guy and Patsy Beston , Professor Jules Petty, Martin Whiteley, Ian Waldron, Sam Thompson ,Hilary King, Shirley Farrar, Mark Dawson, Patrick Griggs, Roger Hart, Peter Bishop, Jody Motterham , Robert Brinton, Joe Lea, Terry Goulden, Nicola Short, Justin and Ellie Ogilvy, Robin Roberts, Robert Keeling, David Garratt, Julia and Danny Rea , Daniel Davies, Pat Dumayne, James Weller, Mike Bowdidge, Marcus Hammond, Sarah Siddique, Dean Prangnell, Monique Jowers, Bradley Elvin , Barry Dyer, Steven Galley, Mark Crouch, Amanda Smalley, Howard Moody, Ann Brewis, Jamie Dixon, Jason Elliott, Lindi Hodgson, Paul Charlton, Richard and Leiamara Marson ,Alex Warren, Jack Mullock, Gary Sharkey, Cheryl Scarrott, Mathew Rowland, Jamie Gould, Sheila Roberts , Chris Jones, Roy Page, Shaun Yale , Stephen Ralph , Maureen McCrery, Paul Smith, Peter Coales, Irena Sccavarraccinni, Dave Vowles , Wendy Pollard, Adam Weightman, Anne Mendelson, Helen Bishop, Dave Pitman, Christine Sunter , Doris and Philip Jennings , Simon Lancaster , Ruth Turner, Nikki Martin, David Stratton, Brenda Sore, Paul Newman, Darren Mason,
Joanne Walsh , Kim Williams, Laura Henry , William Guild, Stuart White, Matthew Finch, Nigel Davis, Brogan Moulden, David Reeve, Esther Stone, Leon Daines, Sharon Hutchinson , Brian Barton, David Sykes, Ann Tindall, Jayce Allatt, Gary Reed, Carol and Geoff Duke, Adam McKale, Lisa Farrow, Tina Wilson-Kallagher, Anne Wilkinson, John Anderson , Robert Rogers , Sue Elsey, Robert Mochrie, Thomas Jennings, Matthew Maines , Keith Hay, Albert Handsley, Peter Laing, Rachel Flynn, David Hancock, John Golder, Chic and Cat Blades, Thomas Anderson , Ashley Jay, Richard Empson , Darren Thomas, Dave Gunn, Shona Cameron, John Irving, Christine Smith , Gary Haggart, Jan miller, mChris Daniels, Vincent Galusha , Ian Hunter, Lisa Butler, Gordon McLean, Steven Clynes, Paul Sinnott , Denise Selway, Stephen Philpot , Sandy Perceval, Dr. Edward Yeargers, Terrance Fletcher, Zoe Paraskevas , Lee Straw, Kelvin Higgins. And all the witnesses who wished anonymity.

Media Credits: Brian Clough (Paisley Gazette), BCIB, Reading Evening Post, Evening Standard, Bucks Free Press, Cambridgeshire Evening News, Runcorn Weekly News, Sunderland Echo, North West Evening Mail, Derby Evening Telegraph, Belper News, Matlock Mercury, Radio Bristol, North Devon Journal, Radio Devon, Dorset Echo, Epping Guardian, Enfield Independent, Gloucestershire Echo, Stroud News & Journal, The Forrester, Cotswold Observer, Hayling Island Today, Andover Advertiser, Hertfordshire Mercury, Essex & Herts News, Daily Mail, Sheerness Times Guardian, Kentish Express Ashford & District, News Shopper, St Albans Observer, Kent Courier, The Leicestershire Mercury, Loughborough Echo, Gina Bolton Radio Leicestershire, Melton Times, Stamford Mercury, Harrow Times, Diss Express, Peterborough Today, Hexham Courant, Northumberland Big Cat Diaries, Shropshire Star, Bridgenorth Journal, Bridgewater Mercury, Bath Chronicle, Somerset County Gazette, Source: Somerset Standard, Yeovil Express, Western Gazette, Weston Mercury, Chard and Ilminster News, This is the West Country, BBC, Heartland Evening News, Stoke Evening Sentinel, Suffolk Evening Star, Your Local Guardian, Eastbourne Today, The Argus, Mid-Sussex Times, Hastings & St Leonards Observer, Wiltshire Times, The Westmoreland Gazette, Swindon Advertiser, Western Morning New, Wilts and Gloucestershire Standard, Doncaster Free Pres The Northern Ech The Press, Spenborough Guardian, Whitby Gazette, Malton and Pickering Mercury, Gazette & Herald, The Scotsman, The Dundee Courier, Banffshire Herald, Kirkintilloch Herald, Fife Courier, West Word, Forres Gazette, Greenock Telegraph, Press and Journal, Big Cat Monitors website, South Wales Evening Post, http://icwales.icnetwork.co.uk, Ballymoney Times, Newry Democrat, The Irish Independent, Tyrone Herald, Tyrone Times, Anita Guidera, Cryptomundo, The State, South Carolina, Advocate & Democrat.

Websites:

Big Cats in Britain www.bigcatsinbritain.org
The Dorset Big Cat Register http://www.dorsetbigcats.org
Rutland & Leicestershire Panther Watch http://www.bigcats.org.uk
Centre for Fortean Zoology: http://www.cfz.org.uk/index.htm
At the Sign of the Black Cat: http://people.bath.ac.uk/liskmj
ECOS: http://www.banc.org.uk
Nature with Attitude: http://www.naturewithattitude.org
MARCA http://www.webace.com.au/~pwest/marca
Derbyshire Big Cats http://www.derbywildcat.bravehost.com
Loren Colemans Cryptomundo: http://www.cryptomundo.com
Beyond Magazine: http://www.beyondthemagazine.com
Big Cat Rescue: http://www.bigcatrescue.org
Dartmoor Hawking School of Falconry: http://www.dartmoorhawking.co.uk
Dick Raynors Big Cats of Loch Ness: http://www.lochnessinvestigation.org/Pumas.html
Beastwatch UK by Chris Mullins: http://www.beastwatch.co.uk
Big Cats On Line: http://www.dialspace.dial.pipex.com/agarman/bco/ver4.htm
Beartracker's Animal Tracks Den: http://www.bear-tracker.com/index.html
Tooth and Claw: http://www.toothandclaw.org.uk
Mysteries Magazine USA: www.mysteriesmagazine.com

THE CENTRE FOR FORTEAN ZOOLOGY

So, what is the Centre for Fortean Zoology?

We are a non profit-making organisation founded in 1992 with the aim of being a clearing house for information, and coordinating research into mystery animals around the world. We also study out of place animals, rare and aberrant animal behaviour, and Zooform Phenomena – little-understood "things" that appear to be animals, but which are in fact nothing of the sort, and not even alive (at least in the way we understand the term).

Why should I join the Centre for Fortean Zoology?

Not only are we the biggest organisation of our type in the world but - or so we like to think - we are the best. We are certainly the only truly global Cryptozoological research organisation, and we carry out our investigations using a strictly scientific set of guidelines. We are expanding all the time and looking to recruit new members to help us in our research into mysterious animals and strange creatures across the globe. Why should you join us? Because, if you are genuinely interested in trying to solve the last great mysteries of Mother Nature, there is nobody better than us with whom to do it.

What do I get if I join the Centre for Fortean Zoology?

For £12 a year, you get a four-issue subscription to our journal *Animals & Men*. Each issue contains 60 pages packed with news, articles, letters, research papers, field reports, and even a gossip column! The magazine is A5 in format with a full colour cover. You also have access to one of the world's largest collections of resource material dealing with cryptozoology and allied disciplines, and people from the CFZ membership regularly take part in fieldwork and expeditions around the world.

How is the Centre for Fortean Zoology organized?

The CFZ is managed by a three-man board of trustees, with a non-profit making trust registered with HM Government Stamp Office. The board of trustees is supported by a Permanent Directorate of full and part-time staff, and advised by a Consultancy Board of specialists - many of whom who are world-renowned experts in their particular field. We have regional representatives across the UK, the USA, and many other parts of the world, and are affiliated with other organisations whose aims and protocols mirror our own.

I am new to the subject, and although I am interested I have little practical knowledge. I don't want to feel out of my depth. What should I do?

Don't worry. We were *all* beginners once. You'll find that the people at the CFZ are friendly and approachable. We have a thriving forum on the website which is the hub of an ever-growing electronic community. You will soon find your feet. Many members of the CFZ Permanent Directorate started off as ordinary members, and now work full time chasing monsters around the world.

I have an idea for a project which isn't on your website. What do I do?

Write to us, e-mail us, or telephone us. The list of future projects on the website is not exhaustive. If you have a good idea for an investigation, please tell us. We may well be able to help.

How do I go on an expedition?

We are always looking for volunteers to join us. If you see a project that interests you, do not hesitate to get in touch with us. Under certain circumstances we can help provide funding for your trip. If you look on the future projects section of the website, you can see some of the projects that we have pencilled in for the next few years.

In 2003 and 2004 we sent three-man expeditions to Sumatra looking for Orang-Pendek - a semi-legendary bipedal ape. The same three went to Mongolia in 2005, and Guyana in 2007. All three members started off merely subscribers to the CFZ magazine.

Next time it could be you!

Project Kerinci, Sumatra - 2003
In search of the bipedal ape Orang Pendek

How is the Centre for Fortean Zoology funded?

We have no magic sources of income. All our funds come from donations, membership fees, works that we do for TV, radio or magazines, and sales of our publications and merchandise. We are always looking for corporate sponsorship, and other sources of revenue. If you have any ideas for fund-raising please let us know. However, unlike other cryptozoological organisations in the past, we do not live in an intellectual ivory tower. We are not afraid to get our hands dirty, and furthermore we are not one of those organisations where the membership have to raise money so that a privileged few can go on expensive foreign trips. Our research teams both in the UK and abroad, consist of a mixture of experienced and inexperienced personnel. We are truly a community, and work on the premise that the benefits of CFZ membership are open to all.

What do you do with the data you gather from your investigations and expeditions?

Reports of our investigations are published on our website as soon as they are available. Preliminary reports are posted within days of the project finishing.

Each year we publish a 200 page yearbook containing research papers and expedition reports too long to be printed in the journal. We freely circulate our information to anybody who asks for it.

Is the CFZ community purely an electronic one?

No. Each year since 2000 we have held our annual convention - the *Weird Weekend* - in Exeter. It is three days of lectures, workshops, and excursions. But most importantly it is a chance for members of the CFZ to meet each other, and to talk with the members of the permanent directorate in a relaxed and informal setting and preferably with a pint of beer in one hand. Starting in 2006 - the *Weird Weekend* has been held in the idyllic rural location of Woolsery in North Devon.

We are hoping to start up some regional groups in both the UK and the US which will have regular meetings, work together on research projects, and maybe have a mini convention of their own.

Since relocating to North Devon in 2005 we have become ever more closely involved with other community organisations, and we hope that this trend will continue. We also work closely with Police Forces across the UK as consultants for animal mutilation cases, and during 2006 we intend to forge closer links with the coastguard and other community services. We want to work closely with those who regularly travel into the Bristol Channel, so that if the recent trend of exotic animal visitors to our coastal waters continues, we can be out there as soon as possible.

We are building a Visitor's Centre in rural North Devon. This will not be open to the general public, but will provide a museum, a library and an educational resource for our members (currently over 400) across the globe. We are also planning a youth organisation which will involve children and young people in our activities.

Apart from having been the only Fortean Zoological organisation in the world to have consistently published material on all aspects of the subject for over a decade, we have achieved the following concrete results:

- Disproved the myth relating to the headless, so-called sea-serpent carcass of Durgan beach in Cornwall, 1975
- Disproved the story of the 1988 puma skull of Lustleigh Cleave
- Carried out the only in-depth research ever into mythos of the Cornish Owlman
- Made the first records of a tropical species of lamprey
- Made the first records of a luminous cave gnat larva in Thailand.
- Discovered a possible new species of British mammal - The Beech Marten.
- In 1994-6 carried out the first archival fortean zoological survey of Hong Kong.
- In the year 2000, CFZ theories where confirmed when an entirely new species of lizard was found resident in Britain.
- Identified the monster of Martin Mere in Lancashire as a giant wels catfish
- Expanded the known range of Armitage's skink in the Gambia by 80%
- Obtained photographic evidence of the remains of Europe's largest known pike
- Carried out the first ever in-depth study of the *ninki-nanka*
- Carried out the first attempt to breed Puerto Rican cave snails in captivity
- Were the first European explorers to visit the `lost valley` in Sumatra
- Made the first video recordings of a new species of scorpion in Guyana, 2007
- Brought the first evidence of red-faced Guyanese pygmies back to the UK in 2007
- Discovered a new colour morph of the rainbow boa

EXPEDITIONS & INVESTIGATIOINS TO DATE INCLUDE

- 1998 Puerto Rico, Florida, Mexico *(Chupacabras)*
- 1999 Nevada *(Bigfoot)*
- 2000 Thailand *(Giant snakes called nagas)*
- 2002 Martin Mere *(Giant catfish)*
- 2002 Cleveland *(Wallaby mutilation)*
- 2003 Bolam Lake *(BHM Reports)*
- 2003 Sumatra *(Orang Pendek)*
- 2003 Texas *(Bigfoot; giant snapping turtles)*
- 2004 Sumatra *(Orang Pendek; cigau, a sabre-toothed cat)*
- 2004 Illinois *(Black panthers; cicada swarm)*
- 2004 Texas *(Mystery blue dog)*
- 2004 Puerto Rico *(Chupacabras; carnivorous cave snails)*
- 2005 Belize *(Affiliate expedition for hairy dwarfs)*
- 2005 Mongolia *(Allghoi Khorkhoi aka Mongolian death worm)*
- 2006 Gambia *(Gambo - Gambian sea monster , Ninki Nanka and Armitage s skink*
- 2006 Llangorse Lake *(Giant pike, giant eels)*
- 2006 Windermere *(Giant eels)*
- 2007 Coniston Water *(Giant eels)*
- 2007 Guyana *(Giant anaconda, didi, water tiger)*

To apply for a **FREE** information pack about the organisation and details of how to join, plus information on current and future projects, expeditions and events.

Send a stamped and addressed envelope to:

**THE CENTRE FOR FORTEAN ZOOLOGY
MYRTLE COTTAGE, WOOLSERY,
BIDEFORD, NORTH DEVON
EX39 5QR.**

or alternatively visit our website at:
www.cfz.org.uk

Other books available from
CFZ PRESS

THE OWLMAN AND OTHERS - 30th Anniversary Edition
Jonathan Downes - ISBN 978-1-905723-02-7

£14.99

EASTER 1976 - Two young girls playing in the churchyard of Mawnan Old Church in southern Cornwall were frightened by what they described as a "nasty bird-man". A series of sightings that has continued to the present day. These grotesque and frightening episodes have fascinated researchers for three decades now, and one man has spent years collecting all the available evidence into a book. To mark the 30th anniversary of these sightings, Jonathan Downes has published a special edition of his book.

DRAGONS - More than a myth?
Richard Freeman - ISBN 0-9512872-9-X

£14.99

First scientific look at dragons since 1884. It looks at dragon legends worldwide, and examines modern sightings of dragon-like creatures, as well as some of the more esoteric theories surrounding dragonkind.

Dragons are discussed from a folkloric, historical and cryptozoological perspective, and Richard Freeman concludes that: "When your parents told you that dragons don't exist - they lied!"

MONSTER HUNTER
Jonathan Downes - ISBN 0-9512872-7-3

£14.99

Jonathan Downes' long-awaited autobiography, *Monster Hunter*...

Written with refreshing candour, it is the extraordinary story of an extraordinary life, in which the author crosses paths with wizards, rock stars, terrorists, and a bewildering array of mythical and not so mythical monsters, and still just about manages to emerge with his sanity intact.......

MONSTER OF THE MERE
Jonathan Downes - ISBN 0-9512872-2-2

£12.50

It all starts on Valentine's Day 2002 when a Lancashire newspaper announces that "Something" has been attacking swans at a nature reserve in Lancashire. Eyewitnesses have reported that a giant unknown creature has been dragging fully grown swans beneath the water at Martin Mere. An intrepid team from the Exeter based Centre for Fortean Zoology, led by the author, make two trips – each of a week – to the lake and its surrounding marshlands. During their investigations they uncover a thrilling and complex web of historical fact and fancy, quasi Fortean occurrences, strange animals and even human sacrifice.

CFZ PRESS, MYRTLE COTTAGE,
WOOLFARDISWORTHY BIDEFORD,
NORTH DEVON, EX39 5QR
w w w . c f z . o r g . u k

Other books available from
CFZ PRESS

ONLY FOOLS AND GOATSUCKERS
Jonathan Downes - ISBN 0-9512872-3-0

£12.50

In January and February 1998 Jonathan Downes and Graham Inglis of the Centre for Fortean Zoology spent three and a half weeks in Puerto Rico, Mexico and Florida, accompanied by a film crew from UK Channel 4 TV. Their aim was to make a documentary about the terrifying chupacabra - a vampiric creature that exists somewhere in the grey area between folklore and reality. This remarkable book tells the gripping, sometimes scary, and often hilariously funny story of how the boys from the CFZ did their best to subvert the medium of contemporary TV documentary making and actually do their job.

WHILE THE CAT'S AWAY
Chris Moiser - ISBN: 0-9512872-1-4

£7.99

Over the past thirty years or so there have been numerous sightings of large exotic cats, including black leopards, pumas and lynx, in the South West of England. Former Rhodesian soldier Sam McCall moved to North Devon and became a farmer and pub owner when Rhodesia became Zimbabwe in 1980. Over the years despite many of his pub regulars having seen the "Beast of Exmoor" Sam wasn't at all sure that it existed. Then a series of happenings made him change his mind. Chris Moiser—a zoologist—is well known for his research into the mystery cats of the westcountry. This is his first novel.

CFZ EXPEDITION REPORT 2006 - GAMBIA
ISBN 1905723032

£12.50

In July 2006, The J.T.Downes memorial Gambia Expedition - a six-person team - Chris Moiser, Richard Freeman, Chris Clarke, Oll Lewis, Lisa Dowley and Suzi Marsh went to the Gambia, West Africa. They went in search of a dragon-like creature, known to the natives as `Ninki Nanka`, which has terrorized the tiny African state for generations, and has reportedly killed people as recently as the 1990s. They also went to dig up part of a beach where an amateur naturalist claims to have buried the carcass of a mysterious fifteen foot sea monster named 'Gambo', and they sought to find the Armitage's Skink (*Chalcides armitagei*) - a tiny lizard first described in 1922 and only rediscovered in 1989. Here, for the first time, is their story.... With an forward by Dr. Karl Shuker and introduction by Jonathan Downes.

BIG CATS IN BRITAIN YEARBOOK 2006
Edited by Mark Fraser - ISBN 978-1905723-01-0

£10.00

Big cats are said to roam the British Isles and Ireland even now as you are sitting and reading this. People from all walks of life encounter these mysterious felines on a daily basis in every nook and cranny of these two countries. Most are jet-black, some are white, some are brown, in fact big cats of every description and colour are seen by some unsuspecting person while on his or her daily business. 'Big Cats in Britain' are the largest and most active group in the British Isles and Ireland This is their first book. It contains a run-down of every known big cat sighting in the UK during 2005, together with essays by various luminaries of the British big cat research community which place the phenomenon into scientific, cultural, and historical perspective.

**CFZ PRESS, MYRTLE COTTAGE,
WOOLSERY, BIDEFORD,
NORTH DEVON, EX39 5QR
w w w . c f z . o r g . u k**

Other books available from
CFZ PRESS

THE SMALLER MYSTERY CARNIVORES OF THE WESTCOUNTRY
Jonathan Downes - ISBN 978-1-905723-05-8

£7.99

Although much has been written in recent years about the mystery big cats which have been reported stalking Westcountry moorlands, little has been written on the subject of the smaller British mystery carnivores. This unique book redresses the balance and examines the current status in the Westcountry of three species thought to be extinct: the Wildcat, the Pine Marten and the Polecat, finding that the truth is far more exciting than the currently held scientific dogma. This book also uncovers evidence suggesting that even more exotic species of small mammal may lurk hitherto unsuspected in the countryside of Devon, Cornwall, Somerset and Dorset.

THE BLACKDOWN MYSTERY
Jonathan Downes - ISBN 978-1-905723-00-3

£7.99

Intrepid members of the CFZ are up to the challenge, and manage to entangle themselves thoroughly in the bizarre trappings of this case. This is the soft underbelly of ufology, rife with unsavoury characters, plenty of drugs and booze." That sums it up quite well, we think. A new edition of the classic 1999 book by legendary fortean author Jonathan Downes. In this remarkable book, Jon weaves a complex tale of conspiracy, anti-conspiracy, quasi-conspiracy and downright lies surrounding an air-crash and alleged UFO incident in Somerset during 1996. However the story is much stranger than that. This excellent and amusing book lifts the lid off much of contemporary forteana and explains far more than it initially promises.

GRANFER'S BIBLE STORIES
John Downes - ISBN 0-9512872-8-1

£7.99

Bible stories in the Devonshire vernacular, each story being told by an old Devon Grandfather - 'Granfer'. These stories are now collected together in a remarkable book presenting selected parts of the Bible as one more-or-less continuous tale in short 'bite sized' stories intended for dipping into or even for bed-time reading. `Granfer` treats the biblical characters as if they were simple country folk living in the next village. Many of the stories are treated with a degree of bucolic humour and kindly irreverence, which not only gives the reader an opportunity to re-evaluate familiar tales in a new light, but do so in both an entertaining and a spiritually uplifting manner.

FRAGRANT HARBOURS DISTANT RIVERS
John Downes - ISBN 0-9512872-5-7

£12.50

Many excellent books have been written about Africa during the second half of the 19th Century, but this one is unique in that it presents the stories of a dozen different people, whose interlinked lives and achievements have as many nuances as any contemporary soap opera. It explains how the events in China and Hong Kong which surrounded the Opium Wars, intimately effected the events in Africa which take up the majority of this book. The author served in the Colonial Service in Nigeria and Hong Kong, during which he found himself following in the footsteps of one of the main characters in this book; Frederick Lugard – the architect of modern Nigeria.

**CFZ PRESS, MYRTLE COTTAGE,
WOOLFARDISWORTHY BIDEFORD,
NORTH DEVON, EX39 5QR
www.cfz.org.uk**

Other books available from
CFZ PRESS

ANIMALS & MEN - Issues 1 - 5 - In the Beginning
Edited by Jonathan Downes - ISBN 0-9512872-6-5

£12.50

At the beginning of the 21st Century monsters still roam the remote, and sometimes not so remote, corners of our planet. It is our job to search for them. The Centre for Fortean Zoology [CFZ] is the only professional, scientific and full-time organisation in the world dedicated to cryptozoology - the study of unknown animals. Since 1992 the CFZ has carried out an unparalleled programme of research and investigation all over the world. We have carried out expeditions to Sumatra (2003 and 2004), Mongolia (2005), Puerto Rico (1998 and 2004), Mexico (1998), Thailand (2000), Florida (1998), Nevada (1999 and 2003), Texas (2003 and 2004), and Illinois (2004). An introductory essay by Jonathan Downes, notes putting each issue into a historical perspective, and a history of the CFZ.

ANIMALS & MEN - Issues 6 - 10 - The Number of the Beast
Edited by Jonathan Downes - ISBN 978-1-905723-06-5

£12.50

At the beginning of the 21st Century monsters still roam the remote, and sometimes not so remote, corners of our planet. It is our job to search for them. The Centre for Fortean Zoology [CFZ] is the only professional, scientific and full-time organisation in the world dedicated to cryptozoology - the study of unknown animals. Since 1992 the CFZ has carried out an unparalleled programme of research and investigation all over the world. We have carried out expeditions to Sumatra (2003 and 2004), Mongolia (2005), Puerto Rico (1998 and 2004), Mexico (1998), Thailand (2000), Florida (1998), Nevada (1999 and 2003), Texas (2003 and 2004), and Illinois (2004). Preface by Mark North and an introductory essay by Jonathan Downes, notes putting each issue into a historical perspective, and a history of the CFZ.

BIG BIRD! Modern Sightings of Flying Monsters

Ken Gerhard - ISBN 978-1-905723-08-9

£7.99

From all over the dusty U.S./Mexican border come hair-raising stories of modern day encounters with winged monsters of immense size and terrifying appearance. Further field sightings of similar creatures are recorded from all around the globe. What lies behind these weird tales? Ken Gerhard is a native Texan, he lives in the homeland of the monster some call 'Big Bird'. Ken's scholarly work is the first of its kind. On the track of the monster, Ken uncovers cases of animal mutilations, attacks on humans and mounting evidence of a stunning zoological discovery ignored by mainstream science. Keep watching the skies!

STRENGTH THROUGH KOI
They saved Hitler's Koi and other stories

Jonathan Downes - ISBN 978-1-905723-04-1

£7.99

Strength through Koi is a book of short stories - some of them true, some of them less so - by noted cryptozoologist and raconteur Jonathan Downes. The stories are all about koi carp, and their interaction with bigfoot, UFOs, and Nazis. Even the late George Harrison makes an appearance. Very funny in parts, this book is highly recommended for anyone with even a passing interest in aquaculture, but should be taken definitely *cum grano salis*.

CFZ PRESS, MYRTLE COTTAGE, WOOLSERY, BIDEFORD, NORTH DEVON, EX39 5QR

Other books available from
CFZ PRESS

BIG CATS IN BRITAIN YEARBOOK 2007
Edited by Mark Fraser - ISBN 978-1-905723-09-6

£12.50

People from all walks of life encounter mysterious felids on a daily basis, in every nook and cranny of the UK. Most are jet-black, some are white, some are brown; big cats of every description and colour are seen by some unsuspecting person while on his or her daily business. 'Big Cats in Britain' are the largest and most active research group in the British Isles and Ireland. This book contains a run-down of every known big cat sighting in the UK during 2006, together with essays by various luminaries of the British big cat research community.

CAT FLAPS! Northern Mystery Cats
Andy Roberts - ISBN 978-1-905723-11-9

£6.99

Of all Britain's mystery beasts, the alien big cats are the most renowned. In recent years the notoriety of these uncatchable, out-of-place predators have eclipsed even the Loch Ness Monster. They slink from the shadows to terrorise a community, and then, as often as not, vanish like ghosts. But now film, photographs, livestock kills, and paw prints show that we can no longer deny the existence of these once-legendary beasts. Here then is a case-study, a true lost classic of Fortean research by one of the country's most respected researchers.

CENTRE FOR FORTEAN ZOOLOGY 2007 YEARBOOK
Edited by Jonathan Downes and Richard Freeman
ISBN 978-1-905723-14-0

£12.50

The Centre For Fortean Zoology Yearbook is a collection of papers and essays too long and detailed for publication in the CFZ Journal *Animals & Men*. With contributions from both well-known researchers, and relative newcomers to the field, the Yearbook provides a forum where new theories can be expounded, and work on little-known cryptids discussed.

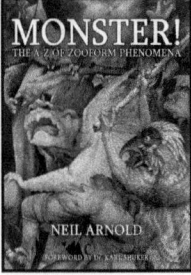

MONSTER! THE A-Z OF ZOOFORM PHENOMENA
Neil Arnold - ISBN 978-1-905723-10-2

£14.99

Zooform Phenomena are the most elusive, and least understood, mystery `animals`. Indeed, they are not animals at all, and are not even animate in the accepted terms of the word. Author and researcher Neil Arnold is to be commended for a groundbreaking piece of work, and has provided the world's first alphabetical listing of zooforms from around the world.

**CFZ PRESS, MYRTLE COTTAGE,
WOOLFARDISWORTHY BIDEFORD,
NORTH DEVON, EX39 5QR
www.cfz.org.uk**

Other books available from
CFZ PRESS

BIG CATS LOOSE IN BRITAIN
Marcus Matthews - ISBN 978-1-905723-12-6

£14.99

Big Cats: Loose in Britain, looks at the body of anecdotal evidence for such creatures: sightings, livestock kills, paw-prints and photographs, and seeks to determine underlying commonalities and threads of evidence. These two strands are repeatedly woven together into a highly readable, yet scientifically compelling, overview of the big cat phenomenon in Britain.

DARK DORSET
TALES OF MYSTERY, WONDER AND TERROR
Robert. J. Newland and Mark. J. North
ISBN 978-1-905723-15-6

£12.50

This extensively illustrated compendium has over 400 tales and references, making this book by far one of the best in its field. Dark Dorset has been thoroughly researched, and includes many new entries and up to date information never before published. The title of the book speaks for itself, and is indeed not for the faint hearted or those easily shocked.

MAN-MONKEY - IN SEARCH OF THE BRITISH BIGFOOT
Nick Redfern - ISBN 978-1-905723-16-4

£9.99

In her 1883 book, *Shropshire Folklore*, Charlotte S. Burne wrote: *'Just before he reached the canal bridge, a strange black creature with great white eyes sprang out of the plantation by the roadside and alighted on his horse's back'*. The creature duly became known as the `Man-Monkey`.

Between 1986 and early 2001, Nick Redfern delved deeply into the mystery of the strange creature of that dark stretch of canal. Now, published for the very first time, are Nick's original interview notes, his files and discoveries; as well as his theories pertaining to what lies at the heart of this diabolical legend.

EXTRAORDINARY ANIMALS REVISITED
Dr Karl Shuker - ISBN 978-1905723171

£14.99

This delightful book is the long-awaited, greatly-expanded new edition of one of Dr Karl Shuker's much-loved early volumes, *Extraordinary Animals Worldwide*. It is a fascinating celebration of what used to be called romantic natural history, examining a dazzling diversity of animal anomalies, creatures of cryptozoology, and all manner of other thought-provoking zoological revelations and continuing controversies down through the ages of wildlife discovery.

**CFZ PRESS, MYRTLE COTTAGE,
WOOLFARDISWORTHY BIDEFORD,
NORTH DEVON, EX39 5QR
www.cfz.org.uk**

Other books available from
CFZ PRESS

DARK DORSET CALENDAR CUSTOMS
Robert J Newland - ISBN 978-1-905723-18-8

£12.50

Much of the intrinsic charm of Dorset folklore is owed to the importance of folk customs. Today only a small amount of these curious and occasionally eccentric customs have survived, while those that still continue have, for many of us, lost their original significance. Why do we eat pancakes on Shrove Tuesday? Why do children dance around the maypole on May Day? Why do we carve pumpkin lanterns at Hallowe'en? All the answers are here! Robert has made an in-depth study of the Dorset country calendar identifying the major feast-days, holidays and celebrations when traditionally such folk customs are practiced.

CENTRE FOR FORTEAN ZOOLOGY 2004 YEARBOOK
Edited by Jonathan Downes and Richard Freeman
ISBN 978-1-905723-14-0

£12.50

The Centre For Fortean Zoology Yearbook is a collection of papers and essays too long and detailed for publication in the CFZ Journal *Animals & Men*. With contributions from both well-known researchers, and relative newcomers to the field, the Yearbook provides a forum where new theories can be expounded, and work on little-known cryptids discussed.

CENTRE FOR FORTEAN ZOOLOGY 2008 YEARBOOK
Edited by Jonathan Downes and Corinna Downes
ISBN 978 -1-905723-19-5

£12.50

The Centre For Fortean Zoology Yearbook is a collection of papers and essays too long and detailed for publication in the CFZ Journal *Animals & Men*. With contributions from both well-known researchers, and relative newcomers to the field, the Yearbook provides a forum where new theories can be expounded, and work on little-known cryptids discussed.

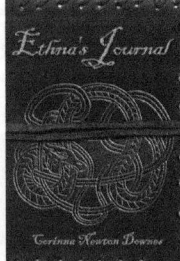

ETHNA'S JOURNAL
Corinna Newton Downes
ISBN 978 -1-905723-21-8

£9.99

Ethna's Journal tells the story of a few months in an alternate Dark Ages, seen through the eyes of Ethna, daughter of Lord Edric. She is an unsophisticated girl from the fortress town of Cragnuth, somewhere in the north of England, who reluctantly gets embroiled in a web of treachery, sorcery and bloody war...

**CFZ PRESS, MYRTLE COTTAGE,
WOOLFARDISWORTHY BIDEFORD,
NORTH DEVON, EX39 5QR
w w w . c f z . o r g . u k**

Other books available from
CFZ PRESS

CFZ PRESS

ANIMALS & MEN - Issues 11 - 15 - The Call of the Wild
Jonathan Downes (Ed) - ISBN 978-1-905723-07-2

£12.50

Since 1994 we have been publishing the world's only dedicated cryptozoology magazine, *Animals & Men*. This volume contains fascimile reprints of issues 11 to 15 and includes articles covering out of place walruses, feathered dinosaurs, possible North American ground sloth survival, the theory of initial bipedalism, mystery whales, mitten crabs in Britain, Barbary lions, out of place animals in Germany, mystery pangolins, the barking beast of Bath, Yorkshire ABCs, Molly the singing oyster, singing mice, the dragons of Yorkshire, singing mice, the bigfoot murders, waspman, British beavers, the migo, Nessie, the weird warbling whatsit of the westcountry, the quagga project and much more...

IN THE WAKE OF BERNARD HEUVELMANS
Michael A Woodley - ISBN 978-1-905723-20-1

£9.99

Everyone is familiar with the nautical maps from the middle ages that were liberally festooned with images of exotic and monstrous animals, but the truth of the matter is that the *idea* of the sea monster is probably as old as humankind itself.

For two hundred years, scientists have been producing speculative classifications of sea serpents, attempting to place them within a zoological framework. This book looks at these successive classification models, and using a new formula produces a sea serpent classification for the 21st Century.

BIG CATS IN BRITAIN YEARBOOK 2008
Edited by Mark Fraser - ISBN 978-1-905723-23-2

People from all walks of life encounter mysterious felids on a daily basis, in every nook and cranny of the UK. Most are jet-black, some are white, some are brown; big cats of every description and colour are seen by some unsuspecting person while on his or her daily business. 'Big Cats in Britain' are the largest and most active research group in the British Isles and Ireland. This book contains a run-down of every known big cat sighting in the UK during 2007, together with essays by various luminaries of the British big cat research community.

**CFZ PRESS, MYRTLE COTTAGE,
WOOLFARDISWORTHY BIDEFORD,
NORTH DEVON, EX39 5QR
w w w . c f z . o r g . u k**

www.ingramcontent.com/pod-product-compliance
Lightning Source LLC
Chambersburg PA
CBHW062156080426
42734CB00010B/1705